AF785995

Studies in Economic Theory

Editors

Charalambos D. Aliprantis
Department of Mathematical Sciences
IUPUI
402 N. Blackford Street
Indianapolis, IN 46202-3216
USA

Nicholas C. Yannelis
Department of Economics
University of Illinois
Champaign, IL 61820
USA

Springer
Berlin
Heidelberg
New York
Barcelona
Budapest
Hong Kong
London
Milan
Paris
Santa Clara
Singapore
Tokyo

Titles in the Series

M. Ali Khan and Nicholas C. Yannelis (Eds.)
Equilibrium Theory in Infinite Dimensional Spaces

Charalambos D. Aliprantis, Kim C. Border and
Wilhelmus A. J. Luxemburg (Eds.)
Positive Operators, Riesz Spaces, and Economics

Donald G. Saari
Geometry of Voting

Charalambos D. Aliprantis and Kim C. Border
Infinite Dimensional Analysis

Jean-Pierre Aubin
Dynamic Economic Theory

Mordecai Kurz (Ed.)
Endogenous Economic Fluctuations

Jean-François Laslier

Tournament Solutions and Majority Voting

With 31 Figures
and 4 Tables

Springer

Jean-François Laslier
Centre National de la Recherche Scientifique
University of Cergy-Pontoise
33, boulevard du Port
F-95011 Cergy Pontoise Cedex, France

ISBN 978-3-642-64561-7 Springer-Verlag Berlin Heidelberg New York

Cataloging-in-Publication Data applied for
Die Deutsche Bibliothek - CIP-Einheitsaufnahme
Laslier, Jean-François: Tournament solutions and majority voting / Jean-François Laslier. - Berlin; Heidelberg; New York; Barcelona; Budapest; Hong Kong; London; Milan; Paris; Santa Clara; Singapore; Tokyo: Springer, 1997
(Studies in economic theory; 7)

ISBN 978-3-642-64561-7 e-ISBN-13: 978-3-642-60805-6
DOI: 10.1007/978-3-642-60805-6

This work is subject to copyright. All rights are reserved, whether the whole or part of the material is concerned, specifically the rights of translation, reprinting, reuse of illustrations, recitation, broadcasting, reproduction on microfilm or in any other way, and storage in data banks. Duplication of this publication or parts thereof is permitted only under the provisions of the German Copyright Law of September 9, 1965, in its current version, and permission for use must always be obtained from Springer-Verlag. Violations are liable for prosecution under the German Copyright Law.

© Springer-Verlag Berlin · Heidelberg 1997
Softcover reprint of the hardcover 1st edition 1997

The use of general descriptive names, registered names, trademarks, etc. in this publication does not imply, even in the absence of a specific statement, that such names are exempt from the relevant protective laws and regulations and therefore free for general use.

Harcover design: Erich Kirchner, Heidelberg

SPIN 10628266 42/2202-5 4 3 2 1 0 – Printed on acid-free paper

To Maryvonne and the family democracy

Acknowledgments

This book is the outcome of a research program that was conducted over the last years in collaboration with Gilbert Laffond, Jean Lainé and Michel Le Breton. Most of the results presented thereafter have been obtained with them and published in co-authored articles, and I am much indebted to them. Thanks are also due to Pavel Chebotarev, Bhaskar Dutta, Olivier Hudry, Slava Levchenkov, Vincent Merlin, Bernard Monjardet, Hervé Moulin, Marc Roubens and Maurice Salles for their encouragement and/or advice at various stage of the preparation of the book. Turning a hieroglyphic manuscript into a decent string of 0s and 1s has been a chore for Sophie Bernier, Christine Leité and Coleen James, and I am most grateful to them. I am beholden to the Centre National de la Recherche Scientifique, the Conservatoire National des Arts et Métiers, and the University of Cergy-Pontoise for their support. The California Institute of Technology, in Pasadena, kindly hosted me while finishing the manuscript, and I enjoyed there an exiting intellectual environment as well as the Millikan library, where most of the bibliographical work was done.

Preface

by Hervé Moulin

" If, when the majorities are found to be cyclical, any elector wish to alter his paper, he may do so;..; but if, when none will make any further alteration, the majorities continue cyclical, there shall be no election. " (C.L. Dodgson, quoted in *Classic of Social Choice*, edited by I. Mc Lean and A. Urken, Michigan University Press 1988). This pessimistic view of the paradox of voting is, in essence, identical to the all too ommon reading of Arrow's theorem as the impossibility of 'rational' collective decisions. In real life, naturally, elections do take place in all weathers, and find a practical way to break cyclical majority preferences.

Two ways out of the impossibility (the most familiar ones among economists) are to restrict the domain of admissible preferences, or to reduce the decisiveness of the social binary comparisons by requiring qualified majorities (or even unanimity) for an unambiguous preference. If domain restriction is relevant to some economic or political models, it has no bite in the fundamental voting problem, where freedom of opinion is a first principle. As for watering down social preferences to a statement of indifference, it fails to meet the need for an unambiguous decision.

Hence the third way out of the impossibility, that retains the sharpness of binary comparisons and computes a fair compromise by some deterministic formula, is for all practical purposes, the only one with a claim to applicability. That this route is traveled more often by mathematicians that by mainstream social scientists is a reflection of its conceptual complexity, not of its lack of relevance. In the voting context, a tournament solution offers a systematic resolution of cyclical majorities, based on the pattern of 'wins and losses' between all pairs of candidates. Beyond the design of voting rules, tournament solutions can also be used to pool the opinions of several experts, untangle individual choices, or rank teams in a sport competition.

This book is the first systematic exposition of the mathematical problem of choice from a tournament. Jean-François Laslier, who with his co-authors has been the leading contributor to the subject for the last six years, offers a comprehensive and technically very deep survey of the many angles from which actual methods of choice can be derived. Thus we visit successively the scoring methods computed from the eigenvalues of a matrix representing the tournament (Chapter 3); the graph theoretical solutions centering on the maximal elements of certain coarsenings of the majority relation (Chapter 5 and 7); an original application of the statistical technique of principal moments (Chapter 4); a new solution obtained from the noncooperative equilibrium of a natural election game (Chapter 6); and an algebraic analysis of the solutions implemented by a sequence of majority votes, a method used by many political bodies (Chapter 8).

Throughout the intricacies of the mathematical analysis of these various methods, the unifying thread is the axiomatic viewpoint, constantly probing the normative justification of the methods by means of a dozen 'tests', and aiming toward a taxonomy to guide our choice of one of them. Thus the author is revealed as a true social scientist, and his work as a splendid example of applied mathematics with a purpose.

Durham, North Carolina, January 1997

Table of Contents

Introduction

Imagine a tennis club in which a competition has been organized in the following way: Each one of the n members of the club plays once against each of the $n-1$ others. Since tennis allows no tie, one has at the end of the competition $n(n-1)/2$ pieces of information of the form "x has won against y". Such data is called a *tournament*[1]. The tournament structure is one of the simplest pairwise comparison structures one can imagine. Given n distinct objects (or "outcomes"), one has a tournament if all the $n(n-1)/2$ pairwise comparisons have been performed and if the result of each of them is without ambiguity either "x is strictly better than y" or "y is strictly better than x" for any two distinct x and y. The main question this book addresses is the following:

Given a tournament, which are the best outcomes ?

---

[1] Or Round Robin. The word Tournament is well established in mathematics, but note that the etymology of the word Round Robin is of some interest for social scientists. Before designing a competition in which every participant meets every other participant, a Round Robin was, in the jargon of 18th century sailors, a petition having the names of the subscribers arranged in a circle so as to disguise the order of signature (Oxford English Dictionary). K. Suzumura told me that Japanese peasants used to do the same thing on an umbrella.

That is to say, we want to know how to *choose* from a tournament. If one member of the tennis club has won against all the other players, then there is no problem in deciding that this person is the tournament winner (such a champion will be called a *Condorcet winner*). But if it is not the case, then the tournament must contain a cycle of the form "x_1 has won against x_2, x_2 has won against x_3, ..., x_k has won against x_1, and nobody has won against all of x_1, x_2, x_3, ... and x_k". In that case, the choice problem "Who are the best players ?" is no longer trivial, and it may be given quite different reasonable answers. Related questions that we shall consider are the questions of *ranking* (linear ordering from best to worst), *scoring* (give points to the outcomes, the best ones getting the most points) and *describing* (make graphical representations of the tournaments).

Apart from pure and applied *Mathematics,* the theory of tournament, and especially the question of choosing from a tournament has both theoretical and practical interest. As a theoretical choice problem, it has been paid attention by researchers involved in *Economics, Social Choice Theory, Voting Theory, Psychology* and *Decision Science.* As a practical choice problem it falls into the fields of *Political Science, Multicriteria Decision Analysis* and *Statistics.* A particular reference must also be made to sport competitions, where school-case examples of tournaments are often taken.

In Psychology, tournaments arise at the individual level for the study of non-transitive preferences. Non-transitive individual preferences are easily observed (Tversky 1969), if not easily explained or taken into account (Ng 1989, Fishburn 1991). Practical tools for the analysis of preferences in Psychology and in Marketing (Batteau, Jacquet-Lagrèze and Monjardet 1981) are based on various pairwise comparisons structures and on tournaments in particular. Pairwise comparisons lead to original problems in Statistics (David 1963, Kendall 1970) and the related literature is abundant (see Davidson and Farquhar (1976) for a bibliography). In Decision

Theory, non-transitive revealed preferences are natural and are usually explained by the existence of contradictory criteria of choice, these multiple criteria being explicit or hidden (Roubens and Vincke 1985, Roy and Bouyssou 1993). Aggregation of multiple transitive binary relations into a single one is also a Social Choice problem when the democratic society takes as choice criteria the multiple opinions (or values) of the citizens. The problem of choosing from a tournament thus has a long and venerable history which can be traced back to childhood of modern democracy. The discussions by Condorcet, Borda and Lhuiller, during the French revolution, on "the best voting method" are a starting point for a vast literature which has both an abstract orientation in Social Choice Theory (Arrow 1951, Fishburn 1973, Moulin 1988) and a more practical, or descriptive, one in Political Science (Dummet 1984, Nurmi 1987, Merrill 1988, Felsenthal and Machover 1992, Brams 1994). The "Condorcet paradox" is the fact that the aggregation, by the majority rule, of transitive preferences need not be transitive. It is a quite frequent phenomenon when no structure is imposed on the individuals' opinions (Fishburn and Gehrlein 1982, Nurmi 1992, Lepelley 1993). The restrictions to be imposed on the individual opinions in order to logically rule out the possibility of a Condorcet paradox are stringent (Inada 1969) and essentially realistic in the case of "One-dimensional geometric voting with single-peaked preferences" (*cf.* Downs 1957). In this framework, where the issue to be voted upon is a number (for instance a tax-rate) or a position on a one-dimensional political axis (left-right), the "Median Voter Theorem" (Black 1948) guarantees the existence of an alternative defeating every other on the basis of pairwise majority voting, and no Condorcet paradoxe arises. But the import of this theorem is limited because, as soon as several dimensions are introduced, cyclical majorities become the rule (McKelvey 1979), and the paradoxe remains even if the various dimensions seem to be independent (Schwartz 1977, Hollard and Le Breton 1995).

Since the problem has been handled by people coming from rather different areas, it is not surprising that the literature on tournaments is most disseminated. Indeed the same concepts appear in different fields under different names, and results are often discovered and re-discovered several times. One of the purposes of this book is to gather and unify these concepts as well as the related results.

There exists a monography devoted to general topics on tournaments, it is John W. Moon's book *Topics on Tournaments,* which was published in 1968. At that time it was possible to write a survey of the existing literature covering the whole theory of tournaments. Due to the growing of interest for the subject and to the exploding number of articles, this is no longer the case, and here we restrict our attention to the choice problem. Our work can be seen as the extension of one chapter of Moon's book entitled "Ranking the Participants in a Tournament", which is only one out of 29.

Organisation of the Book

Mathematically, tournaments can be defined in several equivalent ways. Let X be a finite set.

In Graph Theory: A directed graph (X, A), where X is the set of vertices (or "nodes") and A is the set of arrows (or "arcs"), is called a tournament if it is complete and asymmetric, that is:

$$\forall (x, y) \in X^2, x \neq y \Rightarrow (x,y) \in A \text{ or } (y, x) \in A,$$

$$\forall (x,y) \in X^2, (x,y) \in A \Rightarrow (y,x) \notin A.$$

In Relation Algebra: A binary relation T is a tournament if for any x and y one and only one of the following is true:

$x=y$, xTy or yTx.

In Operation Algebra: An operation $\vee$ on X is a tournament if:

$\forall(x,y) \in X^2, x \vee y \in \{x,y\},$

$\forall(x,y) \in X^2, x \vee y = y \vee x.$

In Matrix Algebra: A square matrix M of order n is a tournament matrix if:

$\forall (i,j) \in \{1, ..., n\}^2, M_{ij} \in \{0,1\}, M_{ii} = 0 \text{ and } (i \neq j \Rightarrow M_{ij} = 1-M_{ji}).$

Throughout this book, tournaments are defined as complete and asymmetric binary relations. The notations and the basic useful facts about tournaments are given in the first chapter. The material of this chapter is not original, with the exception of Theorem 1.4.6. which establishes a link between the two notions of Decomposition and of Regularity.

In the second chapter, a *Tournament Solution* is defined as a way to choose best outcomes (or "winners") in a tournament. Several properties are introduced that a tournament solution may or may not satisfy: Monotonicity, Strong Superset Property, Independence with respect to the Losers, Aïzerman Property, Composition Consistency and Regularity. These properties are somehow intuitively appealing. For instance Monotonicity means that a best outcome must remain so when it is in some sense strengthened. In the remaining of the book, the alternative solutions under consideration are classified according to these properties. In the same chapter is also mentioned Mc Garvey's theorem (Theorem 2.1.2.), which relates the Theory of Tournaments to Majority Voting. Namely, for any tournament, there exists an electoral body such that the result of all majority votes on pairs of possible outcomes yields the given tournament. This theorem, which is an extension of the well-known "Condorcet Paradox" motivates the interest of social scientists for the problem of choosing from a tournament.

Chapter 3 is devoted to the solutions defined by scoring and ranking methods. The simplest solution of "counting the wins" is called the Copeland solution, and we also consider various works attached to the names of C. Berge, T. Wei, V. Levchenkov, P. Slater and others. Several of these methods are familiar to practitioners. Chapter 4 is devoted to original methods which use Principal Component Analysis and Multidimensional Scaling for the description of tournaments ; these methods provide graphical descriptions of tournaments and are intended to be practical ones.

In chapter 5 is introduced the covering relation associated to a tournament. This theoretically important notion is due to N.R. Miller and P.C. Fishburn. It allows for the construction of several solutions, among them B. Dutta's Minimal Covering set. In chapter 6 is defined the Bipartisan solution, which is a refinement of the solutions introduced in the preceding chapter, and which has moreover a nice interpretation in voting theory. Existence of the Bipartisan set is the answer to a problem in political theory as an extension of the celebrated Median Voter theorem. Chapter 7 presents solution concepts due to J. Banks and T. Schwartz. Banks' solution is appealing because it is motivated by works on sophisticated voting on agenda, in particular by R. McKelvey, D. Shepsle and B. Weingast. Schwartz's solution is a refinement of Bank's one, but unfortunately very few things are known about it.

Chapter 8 is devoted to the problem of solving tournaments by the mean of binary trees, essentially following a paper of H. Moulin but using an algebraic presentation. An example of voting on a binary tree is provided by the system of amendments and counter-amendments in use at the Congress.

Throughout the book, attention is paid to the comparison between alternative solutions. For instance, is it the case that the winners according to method *A* are necessarily winners according to method *B* ? (In which case we say that *A* is a refinement of *B*.) Can it be the case that methods *A* and *B* provide entirely contradictory recommendations (that is to say disjoint sets of winners) ? Such questions are sometimes uneasy to answer, and some open problems remain. For instance, is it true that Schwartz's Tournament

Equilibrium set is a refinement of Dutta's Minimal Covering set ? In chapter 9, we try to make quantitative comparisons with respect to Copeland's idea of the number of wins, having in mind that this scoring method is by far the most common one.

For the sake of brevity and completeness, we only consider the case of a complete and asymmetric binary relation. But some of the introduced concepts have natural extension to more general cases. Suppose, for instance, that the comparison between two outcomes x and y gives a quantitative result of the form : out of n individuals, $M(x,y)$ votes for x against y and $M(y,x) = n - M(x,y)$ votes for y against x (or a proportion $\varepsilon(x,y)$ for x and $\varepsilon(y,x)=1-\varepsilon(x,y)$ for y). Call this type of structure a *generalized tournament*. Then one may hope to adapt some of the ideas developed for tournaments to such a more general framework. It turns out that the concepts introduced in chapter 3 and 4 are rather easily adapted. For instance the Copeland method becomes the Borda count and the Slater method becomes the Kemeny rule (median rankings). The Markov method and the multivariate methods also remain valid. For the more sophisticated methods of the next chapters, careful extension of the definitions allow some generalizations. For instance the Bipartisan Set, can also be proved to exist for generalized tournaments, under rather mild assumptions. These extensions are briefly discussed in the last chapter, together with the applications of Tournament Theory to Voting Theory.

The epistemological status of the Tournament Solutions under consideration may vary. Some methods are clearly designed for practical use, such is the case of the scoring, ranking and descriptive methods. Others should be considered as abstract description of existing procedures : such is the case of the Bipartisan Set, if one conceives it as the Nash Equilibrium of a real electoral competition. Finally, other methods are rather pure theoretical objects ; for instance, the Uncovered Set should not be seen as a recommendation for a real choice, but the property for an outcome to be not-covered is a rather appealing quality for that outcome within a given tournament.

The computational needs for the different methods to be applied also vary a lot. It is trivial to find the Copeland winners. Basic linear algebra is sufficient for the Long Path method, the Markov method or the Multidimensional descriptions. Finding a good algorithm for the Slater ranking is a well-documented problem. The Uncovered Set is not difficult to find, and the Bipartisan Set can be computed using algorithms devoted to Nash equilibria of two-player games. Unfortunately, no algorithm has yet been published for finding the Minimal Covering set or the Tournament Equilibrium set of large tournaments. For tournaments of order 10 or more, it is almost impossible to find (in the general case) these sets at hand.

1 - Generalities

In this chapter are gathered most of the general results dealing with the tournament structure that will be needed in the sequel. Basic references on tournaments are Harary and Moser (1966) and Moon (1968). A more recent survey is Reid and Beineke (1978). The last section of the chapter also contains the basic definitions and properties of binary relations which will occasionally be needed.

1.1. Definitions and Notations

Among the various possible definitions, we choose to define a *tournament*, T on a set X as a complete and asymmetric binary relation:

<u>Definition 1.1.1.</u>: Let X be a set and $T \subset X^2$. The relation T is a tournament if and only if:

(i) $\forall x \in X, (x, x) \notin T$

(ii) $\forall (x,y) \in X^2, x \neq y \Rightarrow [(x,y) \in T \text{ or } (y,x) \in T]$

(iii) $\forall (x, y) \in X^2, (x, y) \in T \Rightarrow (y, x) \notin T$

So a tournament defines an oriented graph, whose vertices are the elements of X and the edges are the elements of T. Whenever $(x, y) \in T$ we write xTy or, if there is no ambiguity about the tournament T, $x \rightarrow y$, and we say that x "beats" y. If xTy for all y in $Y \subset X$, we write xTY or $x \rightarrow Y$. We denote by $\mathcal{T}(X)$ the set of tournaments on X. A strict and complete (linear) ordering is a particular case of tournament, we denote by $\mathcal{OS}(X)$ the set of linear orderings of X. The notion of isomorphism for tournaments is defined by the following:

Definition 1.1.2. : Let X and Y be two sets and $T \in \mathcal{T}(X)$ and $U \in \mathcal{T}(Y)$ be two tournaments on X and Y. A mapping φ from X to Y is a *tournament isomorphism* if and only if :

(i) φ is a bijection (one-to-one and onto)

(ii) $\forall (x, x') \in X^2, xTx' \Leftrightarrow \varphi(x) U \varphi(x')$.

We denote by $\sigma(X)$ the permutation group of X and by $Aut(T)$ the group of automorphisms of a tournament $T \in \mathcal{T}(X)$, a sub-group of $\sigma(X)$, which we simply call the *automorphism group* of T.

We are interested in the theory of tournaments, properly speaking, that is to say that all the introduced concepts are stable by tournament isomorphisms : thus all the mentioned properties P will verify that if T satisfies P and U is isomorphic to T, then U also satisfies property P. Given a tournament T, two vertices x and y are *symmetric* if there exists an automorphism φ of T such that $y=\varphi(x)$ (and thus $x=\varphi^{-1}(y)$). A tournament is *vertex-homogeneous* if for every pair (x, y) of vertices, x and y are symmetric. Then all the vertices are equivalent in the sense that a property shared by a vertex is *ipso facto* shared by all the other vertices. (Fried 1970, Reid and Beneke 1978.)

Another very basic notion is the one of *subtournament* of a tournament:

Definition 1.1.3. : Let $Y \subset X$, $T \in \mathcal{T}(X)$ and $U \in \mathcal{T}(Y)$; we say that U is a sub-tournament of T if and only if $U \subset T$. Then we write $U = T/Y$ and we say that U is the *restriction* of T to Y.

1.2. Finite Tournaments

A tournament $T \in \mathcal{T}(X)$ is *finite* if and only if X is finite. In the sequel, we only consider finite tournaments. We denote by $o(T)$, and we call *order* of T the cardinal of X. We denote by $\mathcal{T}_n$ the class of tournaments of order n, and by $\mathcal{T}$ the class of finite tournaments. It is easy to see that, on a set X of cardinal n there are $2^{n(n-1)/2}$ distinct tournaments ; but many of them are isomorphic. It is possible to compute the number of non-isomorphic tournaments on X (see Moon 1968, p. 86). If we consider for instance $n = 8$ points, it is not difficult to draw such graphs at hand, but there are already 6 880 non-isomorphic tournaments. For 10 points, this number goes to 9 733 056. This explains why searching for examples and counter-examples by systematic exploration of the small tournaments gives poor results, even with the help of a computer.

Finite tournaments can be represented by graphs like figure 1.1. This figure depicts the tournament of order 4, T on $X = \{a, b, c, d\}$ such that :

$$aTb,\ aTc,\ bTc,\ bTd,\ cTd \text{ and } dTa.$$

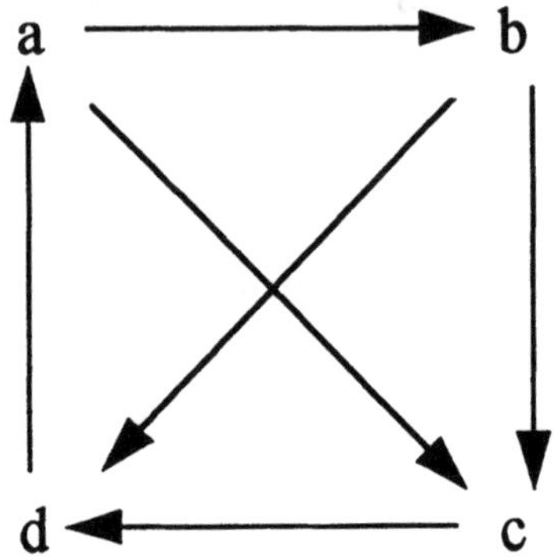

Figure 1.1.

For larger tournaments, this kind of picture becomes intricate so that tournaments are usually depicted using the convention that missing arrows go down. For instance, the tournament in figure 1.1 can be represented by figure 1.2.

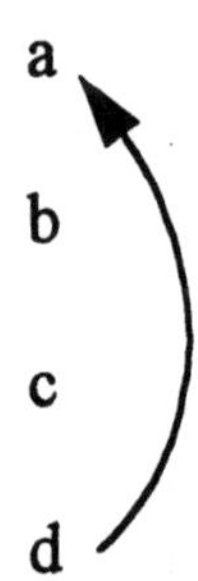

Figure 1.2.

Examples of tournaments :

a) *Linear orderings* : The simplest tournaments are the *transitive tournaments,* that is to say the linear orderings. We shall denote by $<_n$ the tournament defined on $\{1, ..., n\}$ by the usual ordering : $1 <_n 2, ..., <_n n$. Several measures of intransitivity for tournaments have been proposed, in order for instance to quantify the degree of inconsistency shown by a subject's answers to binary comparisons. A measure of inconsistency μ associates to each tournament T a non-negative number $\mu(T)$ such that $\mu(T)=0$ if and only if T is transitive. Kendall and Smith (1940) count the number of 3-cycles for T, that is the number of triples (x, y, z) such that xTy, yTz and zTx. Slater (1961) counts the minimum number of arcs one needs to reverse in order to restore transitivity (see further, chapter 3, section 4). Other measures have been proposed by Ryser (1964), Fulkerson (1965), Bezembinder (1981), Barthélemy (1990) Maas (1993) ; for a review of this literature, see Monsuur and Storken (1996).

b) *Rotational tournaments* : Let us denote by Z_n the set of integers *modulo n* and let Y be a subset of Z_n such that $0 \notin Y$ and for all $y \neq 0$, $y \in Y$ if and only if $-y \notin Y$. Let R_Y be the relation on Z_n defined by: xR_Yx' if and only if $x'-x \in Y$, then R_Y is a tournament, called the rotational tournament of symbol Y. The translations $x \mapsto x+k$ are automorphisms for this tournament, thus the rotational tournaments are vertex-homogeneous. The two following examples are the rotational tournaments that we shall use.

c) *Cyclical tournaments* (*cyclones*) : For any odd integer n, the set $\{1, ..., (n-1)/2\}$ is the symbol of a rotational tournament of order n.

Definition 1.2.1. : Let n be an odd integer, we call *cyclical tournament* of order n, and we denote by C_n the tournament defined on Z_n by :

$$xC_n y \Leftrightarrow y - x \in \{1, ..., \frac{n-1}{2}\}.$$

Dugat (1990) proposed to call this tournament the *cyclone* of order n. Observe that for any alternative x, the restriction of C_n to the set of predecessors of x is transitive of order $(n-1)/2$, and so is the restriction to the set of successors of x. It is easy to see that this property characterizes the cyclones.

d) *Quadratic residues tournaments* : For an integer n, let Q_n be the binary relation on Z_n defined by : $x\ Q_n\ y \Leftrightarrow \exists\, u \in Z_n : y - x = u^2$ and $u \neq 0$.

Proposition 1.2.2. : If n is a prime number equal to 3 *modulo* 4, then Q_n is a tournament.

Proof :

Consider in Z_n the application $\kappa : u \mapsto u^2$. For any u, $\kappa(u)=0$ means that n divides u^2. If n is prime this implies that n divides u and thus $\kappa(u)=0 \Leftrightarrow u=0$. Therefore for no x in Z_n, xQ_nx. Consider now κ on Z_n - $\{0\}$, κ is a multiplicative morphism sending Z_n - $\{0\}$ on some subset Y of Z_n - $\{0\}$. Remark that :

$$\kappa(u) = \kappa(v) \Leftrightarrow (u+v)(u-v)=0 \Leftrightarrow (u=v \text{ or } u=-v).$$

It follows that Y contains $(n-1)/2$ elements. If $-1 \in Y$, $u \in Y \Leftrightarrow -u \in Y$ and Y contains an even number of elements, implying $n \equiv 1 \ (mod\ 4)$. Therefore $n \equiv 3\ (mod\ 4)$ implies that $-1 \notin Y$ and for any $u \in Z_n$ -$\{0\}$, either $u \in Y$ or $-u \in Y$. From $-1 \notin Y$, one deduces that Q_n is asymmetric : if xQ_ny and yQ_nx there exist u and v such that $u^2 = -v^2$ thus $-1 = (u/v)^2 \in Y$. For any x and y, either $x-y \in Y$ or $y-x \in Y$, therefore Q_n is complete. ■

Again, the translations $x \mapsto x + k$ are automorphisms of these tournaments, but a further remarkable property holds, meaning that in such tournaments, all the *edges* are equivalent ("arc-homogeneity"). Let us prove that there is an automorphism sending a particular edge, $(0, 1)$, on any other. This is sufficient to prove that any two edges can be sent one on the other by some automorphism, what is called the "arc-homogeneity".

Proposition 1.2.3. : Let n be a prime integer, $n \equiv 3 \pmod 4$, and $(x, y) \in (Z_n)^2$ with xQ_ny. There is an automorphism $\sigma \in Aut(Q_n)$ such that $x = \sigma(0)$ and $y = \sigma(1)$.

Proof :

Let us write for all $z \in Z_n$: $\sigma(z) = (y - x)\, z + x$, then $\sigma(0) = x$, $\sigma(1) = y$ and for all z and z', $\sigma(z)\, Q_n\, \sigma(z')$ if and only if $(y - x)\,(z - z')$ is a square; but $y - x$ is a square, thus : $\sigma(z)\, Q_n\, \sigma(z') \Leftrightarrow z\, Q_n\, z'$. ■

Quadratic residues tournaments exist for $n \equiv 3 \pmod 4$ not only in Z_n but in the Galois field $GF(p^k)$ for any odd k. Then arc-homogeneity is a characteristic properties of this family of tournaments : a tournament (of order at least 3) is arc-homogeneuous if and only if it is a quadratic residues tournament (Berggren 1972).

1.3. Decomposition

The concept of "decomposition" of a tournament has been introduced in the literature with various names: decomposition, congruence, substitutions, clones, adjacent sets, convex parts, autonomous parts, homogeneous parts etc... See Fried and Laksar (1971), Erdos, Fried, Hajnal and Miller (1972), Moon (1972), Muller, Nesetril and Pelant (1975), Varlet (1976), Habib (1981), Mohring and Radermacher (1984), Astié-Vidal and Mateo (1987), Imrich and Nesetril (1992), Astié-Vidal and Dugat (1993) and Laffond, Lainé and Laslier (1996).

Definition 1.3.1. : Let $T \in \mathcal{T}(X)$, a non-empty subset Y of X is a *component* of T if :

$$\forall (y, y') \in Y^2, \forall x \in X - Y, xTy \Leftrightarrow xTy'.$$

A component Y is *proper* if $Y \neq X$.

Definition 1.3.2. : We call *decomposition* of $T \in \mathcal{T}(X)$ a partition of X in components of T. A decomposition is said *null* if it is reduced to the single set X and *trivial* if it is equal to the set of singletons of X, in the other cases it is said *proper*. A tournament is said to be *decomposable*, or *composed* if it admits a proper decomposition.

The tournaments which are not composed are sometimes called *simple* tournaments. The main interest of this notion of decomposition is that the tournament structure is projected from the set of vertices onto the partition. This allows for the definition of the summary of a tournament *via* a decomposition:

Proposition 1.3.3. : Let $\tilde{X} = \{X_1, \ldots, X_k\}$ be a decomposition of $T \in \mathcal{T}(X)$, $\tilde{X}$ is endowed with a structure of tournament $\tilde{T} \in \mathcal{T}(\tilde{X})$ defined by :

$$\forall X_i, X_j \in \tilde{X}, X_i \; \tilde{T} X_j \Leftrightarrow (X_i \neq X_j \textit{ and } \exists x_i \in X_i : \exists x_j \in X_j : x_i T x_j).$$

Proof : This is an immediate consequence of definition 1.3.1. ■

Definition 1.3.4. : If $\tilde{X}$ is a decomposition of T, the induced tournament $\tilde{T}$ will be called a *summary* of T.

Using a slight abuse of notation, if $\tilde{T}$ is the summary of T via a partition $\{X_1, \ldots, X_n\}$, we sometimes say that $\tilde{T}$ is a tournament on $\{1, \ldots, n\}$. Knowledge of the summary $\tilde{T}$ and of the restrictions T_i is sufficient to recover the initial tournament T. This operation is now defined precisely.

Definition 1.3.5. : Let $\widetilde{T}$ be a tournament of order k defined on $\{1, ..., k\}$ and $T_1, ..., T_k$ be k tournaments defined on k disjoint sets $X_1, ..., X_k$; the (multiple) product of $\widetilde{T}$ by $T_1, ..., T_k$ is the tournament on $X = \bigcup_i X_i$, denoted by : $T = \Pi(\widetilde{T}; T_1, ..., T_k)$ and defined by : $\forall (x,y) \in X^2, x \neq y, x \in X_i, y \in X_j$

$$xTy \Leftrightarrow \begin{cases} i = j \text{ and } xT_iy \\ or \\ i \neq j \text{ and } i\widetilde{T}j. \end{cases}$$

The preceding definitions generalize two notions which appear in the literature:

Definition 1.3.6. : The *product* of the tournaments T' and T'' is the tournament $T = T' \otimes T''$ obtained by replacing each vertex of T' by a copy of T'', that is to say :

$$T = \Pi(T' ; T'', T'', ..., T'').$$

Definition 1.3.7. : We call *reducible* a tournament which possesses a decomposition in *two* tournaments :

$$T = \Pi(\widetilde{T} ; T', T''),$$

since any tournament of order two is (isomorphic to) $<_2$, all the vertices of a component dominate all the vertices of the other component. A tournament which is not reducible is said *irreducible*.

There is no reason for the proper decomposition of a tournament to be unique, whenever such a decomposition exists. Nevertheless it is possible to define a notion of *minimal* decomposition of a tournament.

Definition 1.3.8. : Let $T \in \mathcal{T}(X)$ be a tournament that admits two distinct decompositions $\tilde{X}'$ and $\tilde{X}''$. We say that $\tilde{X}'$ is *coarser* than $\tilde{X}''$ (or $\tilde{X}''$ *finer* than $\tilde{X}'$) if :

$$\forall X''_i \in \tilde{X}'', \exists X'_j \in \tilde{X}' : X''_i \subset X'_j.$$

The finest decomposition is of course the trivial decomposition, and the coarsest is the null decomposition.

Definition 1.3.9.: A decomposition is *minimal* if its only coarser decomposition is the null decomposition.

Remark 1.3.10.: There may be several minimal decompositions, like in the following example : Take $<_3$ the natural linear order on *{1, 2, 3}*, and $\tilde{X}' = \{\{1\}, \{2, 3\}\}$ and $\tilde{X}'' = \{\{1,2\}, \{3\}\}$. Nevertheless, one can prove the following theorem :

Theorem 1.3.11.: If $T \in \mathcal{T}$ is irreducible, then T has a unique minimal decomposition.

Proof :

We need two claims:

claim 1: Let Y and Y' be two components of a tournament T, if $Y \cap Y' \neq \emptyset$, then $Y \cup Y'$ is a component of T.

proof of claim 1 : Let $x \in Y \cap Y'$ and $z \in X- (Y \cup Y')$. If $y \in Y$, $zTy \Leftrightarrow zTx$ because Y is a component, and if $y \in Y'$, $zTy' \Leftrightarrow zTx$ because Y' is a component.

claim 2: Let Y and Y' be two proper components of T with $Y \neq Y'$ and $Y \cup Y' = X$ then T is reducible.

proof of claim 2 : If $Y \cap Y' = \emptyset$ then T is reducible by definition 1.3.7. If $Y \cap Y' \neq \emptyset$ then $Y - Y' \neq \emptyset$ and $Y' - Y \neq \emptyset$. Let $x \in Y \cap Y'$, $y \in Y - Y'$ and z, $z' \in Y' - Y$. Because Y is a component, $yTz \Leftrightarrow xTz$; because Y' is a component, $yTz \Leftrightarrow xTz'$; and because Y is a component, $yTz' \Leftrightarrow xTz'$. Therefore we obtain $xTz \Leftrightarrow xTz'$ and $yTz' \Leftrightarrow xTz'$, which proves that $Y' - Y$ is a component. Thus $\{Y, Y' - Y\}$ is a decomposition and T is reducible.

end of the proof of the theorem : For any $x \in X$ let $Y(x)$ be the union of the proper components to which x belongs. One can deduce from claim 1 that $Y(x)$ is a component of the tournament T, and from claim 2 that $Y(x) \neq X$ because T is irreducible. The family $\{Y(x), x \in X\}$ covers X and verifies :

$$\forall (x, x') \in X^2,\ Y(x) \neq Y(x') \Rightarrow Y(x) \cap Y(x') = \emptyset.$$

Hence it is possible to extract from this family a partition of X, which is of course the minimal decomposition of T. ■

The reducible tournaments also have a most appealing decomposition. Before introducing it, one more definition is needed.

Definition 1.3.12.: Let T be a tournament. A *scaling* decomposition of T is a decomposition of T with a transitive summary : $T = \Pi(<_n ; T_1, ..., T_n)$.

Clearly, reducible tournaments (of order larger than 2) do admit proper scaling decompositions; the converse is also true, as the reader will easily check. Scaling decompositions are in general not unique, but they can be used in order to define a standard way to summarize the reducible tournaments. This is the purpose of the following theorem.

<u>Theorem 1.3.13.</u> If $T \in \mathcal{T}$ is reducible, then it admits a unique proper scaling decomposition such that each component is irreducible.

Proof :

By definition 1.3.7. $T \in \mathcal{T}(X)$ admits two components X' and X'' such that all the vertices of X' dominates all the vertices of X'', if $T' = T/X'$ or $T'' = T/X''$ are reducible, one can iterate the process until one gets a scaling decomposition such that each component is irreducible. This establishes the existence part of the theorem. Suppose there are two such decompositions. For each $x \in X$, let $Y(x)$ and $Z(x)$ be the two components to which x belongs, according to the two scaling decompositions. Then $Y(x) \cap Z(x)$ is a component of T, thus of $T/Y(x)$. If $Y(x)-Z(x)$ is non-empty, it is also a component of T and thus of $T/Y(x)$, but then $T/Y(x)$ is reducible, hence $Y(x)-Z(x)$ is empty. This is enough to prove that $Y(x)=Z(x)$ and the theorem. ■

The scaling decomposition into irreducible components which is presented in this theorem is also the finest scaling decomposition. The two preceding theorems give a way to iteratively decompose any tournament: if the tournament is reducible, one takes the finest scaling decomposition and if it is reducible, one takes the minimal decomposition. Then the operation is done again and again for each component. At the end, all the remaining subtournaments are simple.

Irreducible tournaments also possess the property that any vertex can (by a path of some length) be connected to any other vertex. A graph sharing this property is said *strongly connected,* or simply *strong*. The equivalence of these two properties for tournaments is a classical result, so that, speaking of tournaments, the two notions are identical : a tournament is irreducible if and only if it is strong. More generally, the *Top-Cycle* of *T* is the set of points that indirectly beat all the other points :

Definition 1.3.14.: Let $T \in \mathcal{T}(X)$, we call Top-Cycle of *T* the set

$$TC(T) = \{x \in X : \forall y \in X, \exists k > 0 : \exists (z_1, \ldots, z_k) \in X^k : z_1 = x, z_k = y, \text{ and } 1 \leq i < j \leq k \Rightarrow z_i T z_j\};$$

we say that *T* is *strong* if $TC(T)=X$.

The Top-Cycle provides the first example of what will be called in the following chapter a tournament solution. It is a rather basic graph-theoretical notion (Camion 1959), which was lately considered by social scientists (Good 1971, Schwartz 1972, Fishburn 1974, Bordes 1976, Deb 1977) under the name "Condorcet set". The Top-Cycle is generally a large subset of X, and all the other solutions which will be introduced later choose subsets of the Top-Cycle.

Proposition 1.3.15.: For any tournament *T*, $TC(T)$ is a component of *T*, whose all points beat all the points outside $TC(T)$.

Proof :

In effect, if $x \in TC(T)$ and yTx then $y \in TC(T)$ (this property is called "Condorcet transitivity"). ■

Proposition 1.3.16.: Let $T \in \mathcal{T}(X)$, $TC(T)$ is the first component of T according to its finest scaling decomposition and $TC(T)=X$ (that is : T is strong) if and only if T is irreducible.

Proof :

If $TC(T) \neq X$, the preceding proposition proves that T is reducible. On the other hand, if T is reducible, let Y be the first component of T according to its finest scaling decomposition, then Y beats all the other components for the transitive summary, thus all the points of Y (directly) beats all the points outside Y. A tournament being asymmetric, no point outside Y beats a point of Y. Hence $TC(T) \subset Y$. Since T/Y is irreducible, one in fact has that $TC(T) = Y$. ■

Some more properties of the Top-Cycle will be stated in chapter 8. The next proposition will establish a connection between the two notions of decomposition and of morphism of tournaments. One definition is previously needed :

Definition 1.3.17.: Let X and Y be two sets and $T \in \mathcal{T}(X)$ and $U \in \mathcal{T}(Y)$ be two tournaments on X and Y. A mapping φ from X to Y is a *tournament morphism* if and only if:

(i) φ is a surjection (onto)

(ii) $\forall\ (x, x') \in X^2,\ xTx' \Leftrightarrow \varphi(x)\ U\ \varphi(x')$.

If there exists a morphism from T to U, one says that U is a *morphic image* of T. Bijective morphisms are isomorphisms (definition 1.1.2.) and, unless this could induce some confusion, we usually do not distinguish between isomorphic tournaments. But when is a tournament U a morphic image of T ? We now prove that this only happens if U is a summary of T.

<u>Proposition 1.3.18.</u>: Let $T \in \mathcal{T}(X)$. If there exists a tournament morphism $\varphi : X \to Y$ sending T on $U \in \mathcal{T}(Y)$, then there exist $T^*; T_1, ..., T_k$ such that $T = \Pi\ (T^*; T_1, ..., T_k)$, with T^* isomorphic to U. Conversely, if T^* is a summary of T then there exists a tournament morphism sending T on T^*.

Proof:

Let φ be a morphism from T to U denote φ^{-1} the inverse of φ. Because φ is surjective, $X^* = \{\varphi^{-1}(y) : y \in Y\}$ is a partition of X. Let us prove that this partition is a decomposition. Let x, x', x'' in X with $y = \varphi(x) = \varphi(x') \neq \varphi(x'')$, then: $xTx'' \Leftrightarrow \varphi(x)\ U\ \varphi(x'') \Leftrightarrow \varphi(x')\ U\ \varphi(x'') \Leftrightarrow x'Tx''$, hence $\varphi^{-1}(y)$ is a component of T and X^* is a decomposition for T. Let T^* be defined on T by $\varphi^{-1}(y)\ T^*\ \varphi^{-1}(y') \Leftrightarrow yUy'$ and T_i on $X_i \in X^*$ be defined by $T_i = T/X_i$, then one has: $T = \Pi\ (T^*; T_1, ..., T_k)$ with T^* isomorphic to U. Conversely, the requested morphism is the projection sending any point $x \in X$ on its component for T^*. ■

1.4. Regularity

<u>Definition 1.4.1.</u>: Let $T \in \mathcal{T}(X)$ and $x \in X$, let $T^+(x) = \{y \in X : xTy\}$ and $T^-(x) = \{y \in X : yTx\}$. The elements of $T^+(x)$ are called the *successors* of x and those of $T^-(x)$ the *predecessors*, or *winners* of x.

<u>Definition 1.4.2.</u>: The number of successors of x is called the *Copeland score* of x : $s(x) = \#T^+(x)$. We shall sometimes write $s^+(x)$ instead of $s(x)$, and $s^-(x) = \#T^-(x) = o(T) - s(x) - 1$.

When no further precision is needed, the *score* of a vertex is to be understood as its Copeland score ; nevertheless, since we shall be interested in several different scoring methods (chapter 3), we use a specific name for this most common notion of score. The reference is Copeland (1951).

Definition 1.4.3. : A tournament is *regular* if all the points have the same Copeland score.

The sum of the scores in a tournament of order n is equal to $\frac{n(n-1)}{2}$. For a regular tournament, the score of any point is $\frac{n-1}{2}$, and thus the order of a regular tournament is odd. Vertex-homogenous tournaments are regular. A regular tournament is irreducible. Cyclical tournaments as well as quadratic residues tournaments are examples of regular tournaments.

We shall be interested by the regular composed tournaments. Regular composed tournaments have interesting properties, some of them are mentioned now. The results here gathered will be useful in chapter 5, especially the powerful theorem 1.4.6.

Proposition 1.4.4. : If T is regular and if Y is a component of T, then T / Y is regular.

Proof :

Let $y \in Y$, $s(y) = \#\{x \in X : yTx \text{ and } x \notin Y\} + \#\{x \in Y : yTx\}$, the first term being independent of y in Y, the second term must be too. ■

Proposition 1.4.5. : Any summary of a regular tournament has odd order.

Proof :

The order of the tournament is equal to the sum of the orders of the components, all of them being odd ; thus the summation must be made over an odd number of terms. ■

<u>Theorem 1.4.6.</u> : Let T be a regular tournament and $\widetilde{T}$ be a summary of T, with $T = \Pi(\widetilde{T} ; T_1, ..., T_k)$. Then $\widetilde{T}$ is regular if and only if all the components T_i have the same order.

Proof :

If all the components have the same order, it is straightforward to establish that $\widetilde{T}$ is regular. The converse is obtained in two steps :

First step : Each T_i is regular, of order $2d_i + 1$; denote $d = Min\{d_i : i = 1... k\}$. Let T'_i be the cyclical tournament of order $2(d_i - d) + 1$ and let T' be the composed tournament, with $\widetilde{T}$ as a summary : $T' = \Pi(\widetilde{T} ; T'_1, ..., T'_k)$. It is easily checked that T' is regular, moreover, this decomposition of T' involves at least one component of order 1.

Second step : It is proven that the decomposition of T' is the trivial decomposition. Denote by m the score in T' and by $d'_i = d_i - d$ the score in T'_i. One has for all i :

$$m = d'_i + \sum_{i \to j}(2d'_j + 1) .$$

We may suppose $d_1 = 0$, one then has, letting $k' = \dfrac{k-1}{2}$:

$$\forall i \in \{2, ..., k\}, \qquad m = d'_i + 2 \sum_{i \to j} d'_j + k' \qquad (1)$$

$$m = 2 \sum_{i \to j} d'_j \qquad (2)$$

Let then b be the largest integer such that 2^b divides d'_j for all j, b is well defined, except if all the d'_j are equal to zero ; but it follows from (2) that $m - k'$ can be divided by 2^{b+1} and thus from (1) that, for any i, it is the same for d'_i. This proves that all the d'_i are equal to zero, hence the theorem. ■

Although it is not difficult to find examples of such regular composed tournaments (take the product of two regular tournaments), the reader may wonder whether such tournaments exist at any (odd) order. The answer to this question is given by the following proposition.

Proposition 1.4.7. : For any integer $N>0$, there exists a regular composed tournament of order N if and only if N is odd and $N \geq 9$.

Proof :

Let us first establish that there exists no regular composed tournament of order 1, 3, 5 or 7:

According to the definition, a tournament T on X is said to be composed if it admits a proper component. Let Y be such a component, $\#X>\#Y>1$; from proposition 1.4.4. T/Y is regular, thus from proposition 1.4.5. $\#Y \geq 3$ and $\#X \geq 5$. If there exists $x \in X - Y$ such that xTY then $s(x) \geq 3$ and thus $o(T) \geq 7$, if there is no such x, then for $y \in Y$, $s(y) \geq 3$ and again $o(T) \geq 7$. So suppose that $o(T)=7$. Then the score in T is 3 and it is easily seen that $\#Y=3$, thus each vertex of Y dominates two vertices outside Y and is dominated by two others.

But Y being a component of T, this implies that there exist two vertices that dominate Y and thus there exists a vertex whose score is at least 4, a contradiction.

As a second step, let us prove that if N is an odd integer with $N=9$ or $N\geq 13$ there exists a regular composed tournament of order N:

Let n be an integer. Consider on $\{1, ..., 2n+1\}$ the cyclone of order $2n+1$, C_{2n+1}. Now let m be an integer, with $m\leq n$, consider three disjoints sets X, Y and Z with $\#X= \#Y=2n+1$ and $\#Z=2m+1$ and write: $X=\{x_1, ..., x_{2n+1}\}$, $Y=\{y_1, ..., y_{2n+1}\}$ and $Z=\{z_1, ..., z_{2m+1}\}$. We built a tournament $T(n, m)$ on the set of vertices $X\cup Y\cup Z$ in the following way :

- Inside X : $x_i T(n,m) x_j$ iff $iC_n j$.
- Inside Y : $y_i T(n, m) y_j$ iff $iC_n j$.
- Inside Z : $z_i T(n, m) z_j$ iff $iC_m j$.
- Between X and Y : $x_i T(n,m) y_j$ iff $j-i\in\{1, ..., n+m\}$.
- Between X and Z : for all x_i, $ZT(n, m)x_i$.
- Between Y and Z : for all y_i, $y_i T(n,m) Z$.

The reader will verify that $T(n, m)$ is a regular tournament of order $4n+2m+3$. The subset Z is a component of this tournament, thus $T(n, m)$ is composed if $m\geq 1$. So we have established the existence of a regular composed tournament of order N for any $N=4n+2m+3$, with $1\leq m\leq n$. The assertion follows immediately.

The last remaining value is $N=11$. The example below (figure 1.3) shows a regular composed tournament of order 11. According to a useful tradition, non-depicted arrows go from top to bottom. The set $\{b_1, b_2, b_3\}$ is a component for this tournament, so that there is no need to draw all the arrows between the points b_i and the other points. ■

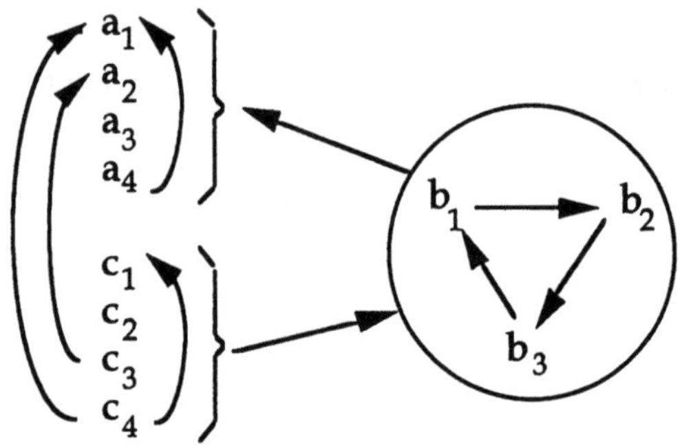

Figure 1.3

5. Useful Notions about General Binary Relations

In this section, we gather some definitions and easy results about binary relations. These notations will be occasionally needed but the section can be skipped in first reading. Let $\mathcal{R}(X)$ denote the set of binary relations on X. We always suppose that X is non-empty and finite. An $R \in \mathcal{R}(X)$ can be seen as a subset of X^2: $R \subset X^2$ and $(x, y) \in R$ is just another way to write xRy. For a subset $Y \subset X$, the *restriction* of R to Y, denoted R/Y is the binary relation $R \cap Y^2 \in \mathcal{R}(Y)$.

Definition 1.5.1.: Let $R \in \mathcal{R}(X)$ and $x \in X$. We say that x is a *maximal element* of R if there exists no $y \neq x$ such that yRx. The set of maximal elements of R is denoted $Max(R)$.

Definition 1.5.2.: Let $\vec{x} = (x_0, x_1, ..., x_n) \in X^{n+1}$. We say that $\vec{x}$ is a *cycle* for $R \in \mathcal{R}(X)$ if $x_0 = x_n$ and for all $i \in \{1, ..., n\}$, $x_{i-1} R x_i$.

We sometimes say that a (non-empty) subset Y of X *forms a cycle* if it is possible to rank the elements of Y in order to obtain a cycle. Observe that a singleton $\{x\}$ forms a cycle if and only if xRx.

<u>Definition 1.5.3.:</u> A binary relation $R \in \mathcal{R}(X)$ is *acyclic* if no subset of X with several elements forms a cycle for R.

Clearly, for $Y \subset X$, if $R \in \mathcal{R}(X)$ is acyclic, then R/Y is acyclic, but the converse is false. It is also easy to see that, for tournaments, acyclicity is a synonym for transitivity. Dealing with finite sets, acyclicity is the appropriate notion for the existence of maximal elements:

<u>Proposition 1.5.4.:</u> Let $R \in \mathcal{R}(X)$, if R is acyclic then $Max(R) \neq \emptyset$ and moreover, if

$$\forall\, Y \subset X,\ [Y \neq \emptyset \Rightarrow Max\ (R/Y) \neq \emptyset]$$

then R is acyclic.

(The proofs of this proposition and of the next ones, are left to the reader.)

<u>Definition 1.5.5.:</u> For $R \in \mathcal{R}(X)$ a binary relation on X and $Y \subset X$, we say that Y is *retentive* for R if Y is non-empty and if there exists no $x \in X\text{-}Y$ and $y \in Y$ such that xRy.

If one interprets xRy as "it is possible to go from y to x" then Y is retentive means that "once in Y it is impossible to get out of Y". We shall define the Top-Set of R as the smallest set in which "one certainly arrives, but can not get out".

<u>Remark 1.5.6.:</u> Let $R \in \mathcal{R}(X)$ and let Y be retentive for R. Let Z be a non-empty subset of Y. Then Z is retentive for R if and only if Z is retentive for R/Y, the restriction of R to Y.

This remark allows for the following definition:

<u>Definition 1.5.7.:</u> Let $R \in \mathcal{R}(X)$. A subset Y of X is *minimal retentive* for R if Y is retentive for R and contains no strict subset Z retentive for R.

Observe that, trivially, X is retentive. Because we only consider finite sets, minimal retentive sets exist.

<u>Remark 1.5.8.:</u> Let $R \in \mathcal{R}(X)$, if Y_1 and Y_2 are two retentive subsets for R such that $Y_1 \cap Y_2 \neq \emptyset$ then $Y_1 \cap Y_2$ is retentive for R.

One can deduce from this observation:

<u>Remark 1.5.9.:</u> Two minimal retentive subsets are either equal or disjoint.

<u>Definition 1.5.10.:</u> The *Top-Set* of R is the union of the minimal retentive subsets of R. It is denoted $TS(R)$. It is non-empty.

<u>Proposition 1.5.11.:</u> The Top-Set of $R \in \mathcal{R}(X)$ is the smallest subset Y of X such that:

- Y is retentive for R and
- for any $x \in X - Y$, there exist $x_0, x_1, \ldots, x_n \in X$ such that : $x_0 = x$, $x_n \in Y$ and, for $i = 1, \ldots, n$, $x_i \, R \, x_{i-1}$.

The Top-Set generalizes the notion of Maximal elements. In particular, for acyclic relations, where maximal elements exist, the two notions are identical. The Top-Set also generalizes the notion of Top-Cycle, which was only defined for tournaments.

Remark 1.5.12.: Let $x \in X$, and $R \in \mathcal{R}(X)$, then $\{x\}$ is retentive if and only if x is maximal.

Remark 1.5.13.: $Max(R) \subset TS(R)$, and if R is acyclic then $Max(R) = TS(R)$.

Proposition 1.5.14.: The Top-Set of a tournament is its Top-Cycle.

Proof:

Let $T \in \mathcal{T}(X)$. Denote $Y=TC(T)$. For any $y \in Y$ and $x \in X\text{-}Y$, yTx (proposition 1.3.15.), therefore Y is retentive for T and any retentive subset is contained in Y. Let Z be retentive, and suppose that Z is a strict subset of Y. Take $y \in Y\text{-}Z$ and $z \in Z$, by definition 1.3.14. there exists $z_0, z_1, \ldots, z_k$ such that $z_0 = y$, $z_k = z$ and for all $i \in \{1, \ldots, k\}$, $z_{i-1}\ T\ z_i$. Let j be the last indice such that $z_j \notin Z$, then $z_{j+1} \in Z$ and $z_j T z_{j+1}$, hence Z is not retentive for T. ■

2 - Tournament Solutions

If a binary relation R on X is a linear order then the problem of choosing according to R is most simple : the function which associates $Max(R/Y)$ to Y is well-defined, single-valued and satisfies all the rationality requirements that one would intuitively like to be satisfied by a "reasonable way of choosing". But, as soon as R does not meet all the linear ordering requirements, the choice problem is no longer trivial and one needs a *Theory of Choice.* Here we develop such a theory for the specific case where R is a tournament. The theory contains the definition of a choice function for tournament ("tournament solution") and the definition and study of some properties a solution may or may not satisfy. Each one of these properties is "intuitively appealing" and would be trivially satisfied if one were to redefine it for linear orderings. Therefore the chapter begins by borrowing to the theory of voting an argument why, dropping transitivity, there is no intermediate case to be considered between linear orderings and general tournaments. A general theory of choice is considered by Aïzerman and Aleskerov (1995), while Moulin (1985) provides a summary of the main results. Surveys for the tournament case are Moulin (1986), Laffond, Laslier and Le Breton (1995a) and Laslier (1996b).

2.1. Majority Voting and Tournaments

A *complete order* on X is a binary relation R on X which is reflexive (xRx for all x) complete (xRy or yRx for all $x{\neq}y$) and transitive (xRy and yRz implies xRz). Complete orders usually model individual preferences. Mac Garvey (1953) has observed that any tournament can be seen as the recording of the results of majority voting between pairs of alternatives, in a sufficiently large society. Let I be a finite set of individuals and X be a finite set of alternatives. The preferences of an individual $i \in I$ are represented by a complete order P_i defined on X. Let $OC(X)$ denote the set of complete orders on X. The vector of the preferences of all the individuals of the population is called the preference profile of the society:

Definition 2.1.1. : A *preference profile* is an element $P = (P_i)_{i \in I}$ of $[OC(X)]^I$.

Pairwise majority voting can now be defined, it associates to each preference profile a binary relation.

Definition 2.1.2. : The result of majority voting is the binary relation $M(P)$ on X such that for any x and y in X:

$$x\, M(P)\, y \Leftrightarrow \#\{i \in I : xP_iy\} > \#\{i \in I : yP_ix\}.$$

If the initial preferences P_i are strict (that is : for no $x{\neq}y$, $x\,P_i\,y$ and $y\,P_i\,x$) and if the number of individuals is odd, the relation $M(P)$ is a tournament. Conversely, one has the proposition :

<u>Proposition 2.1.3.</u> : **Let T be a tournament of order n, there exists a preference profile P on a set I of $n(n - 1)$ individuals such that $T = M(P)$.**

Proof :

One can suppose that T is defined on $X = \{1, \ldots, n\}$. To each of the $\frac{n(n-1)}{2}$ edges (x, y) of T one associates two individuals, the first one has her preference P_{xy} defined by :

$$x \, P_{xy} \, y \, P_{xy} \, 1 \, P_{xy} \, 2 \, P_{xy} \ldots P_{xy} \, n$$

and the second one has her preference P'_{xy} defined by :

$$n \, P'_{xy} \, (n - 1) \, P'_{xy} \ldots P'_{xy} \, 1 \, P'_{xy} \, x \, P'_{xy} \, y.$$

Straightforward verification shows that a strict majority of these $n(n - 1)$ individuals prefers x to y. ■

In other words, each arrow $x \rightarrow y$ of the tournament expresses the result of the comparison between x and y on the basis of the voting of the population. The proposition is valid for complete binary relations because asymmetry of the tournament plays no role in the proof. We know the order of magnitude of the minimal size of the population needed to generate all the tournaments of order n ; this number goes to infinity like $\frac{n}{Log\, n}$ (Stearns 1959, Erdös and Moser 1964, Moon 1968, Deb 1976). If individual preferences are supposed to be separable, any separable tournament may arise (Hollard and LeBreton 1995).

2.2. Solution Concepts

A *social choice correspondence* is a way to choose a subset of alternatives from a preference profile. Fishburn (1977) calls social choice correspondences of class *C1* those social choice correspondences which can be defined by pairwise majority voting. When (and this is the case in which we are interested) the objective of the social choice is to define the "best" alternatives for the population, a minimal requirement for a social choice correspondence of class *C1* is the *Condorcet principle.* Denote by $\mathcal{P}(X)$ the set of subsets of X.

<u>Definition 2.2.1.</u> : Let $F : [OC(X)]^I \to \mathcal{P}(X)$ be a social choice correspondence. One says that F satisfies the Condorcet principle if for all preference profiles P and all alternatives x :

$$[\forall y \in X, xM(P)y] \Rightarrow F(P) = \{x\}.$$

This definition is also valuable for tournaments, independently of the preference profiles from which they can be issued. We will simply call tournament *solution* any correspondence S which, to a tournament $T \in \mathcal{T}(X)$ associates a non-empty subset $S(T)$ of X, stable by tournament isomorphism and which satisfies the Condorcet principle.

<u>Definition 2.2.2.</u> : Let $T \in \mathcal{T}(X)$, the set of Condorcet winners of T is:

$$Cond(T) = \{x \in X : \forall y \in X, y \neq x \Rightarrow xTy\}.$$

Clearly, $Cond(T)$ is either empty either a singleton.

<u>Definition 2.2.3.</u> : A *tournament solution*, S, associates to any tournament $T \in \mathcal{T}(X)$ a subset $S(T) \subset X$ and satisfies:

(i) $\forall T \in \mathcal{T}, S(T) \neq \emptyset$

(ii) for any tournament isomorphism φ, $\varphi \circ S = S \circ \varphi$

(iii) $\forall T \in \mathcal{T}(X), Cond(T) \neq \emptyset \Rightarrow S(T) = Cond(T)$.

We shall sometimes call S a "solution concept" and $S(T)$ the solution of T with respect to the concept S. When there can be no ambiguity about the tournament under consideration, we shall write $S(X)$ instead of $S(T)$. The elements of $S(T)$ will sometimes be referred to as the *winners* of T (with respect to solution S), or the *S-winners*; the other elements of X are the *S-losers*. It is easily checked that TC, the correspondence which associates to any tournament its Top-Cycle is a tournament solution. Observe that, with our definition, the Condorcet correspondence, *Cond*, is not a tournament solution because it may be empty.

If S_1 and S_2 are two tournament solutions, we shall denote by $S_1 \circ S_2$ the tournament solution defined by $S_1 \circ S_2(T) = S_1(T \,/\, S_2(T)) = S_1\ (S_2(T))$. Also, for concept S, we use the notations : $S^1 = S$, $S^{k+1} = S \circ S^k$, and $S^\infty = \bigcap_{k=1}^{\infty} S^k$. Given that all the tournaments under consideration are finite, S^∞ is again a tournament solution.

For comparing solutions, we use the following notations : let S_1 and S_2 be two solutions :

- $S_1 \subset S_2$ means that for all tournaments T, $S_1(T)$ is a subset (strict or not) of $S_2(T)$. One then says that S_1 is finer, or more selective, than S_2, or that S_1 is a refinement of S_2.

- $S_1 \emptyset S_2$ means that there exists at least one tournament T such that $S_1(T) \cap S_2(T) = \emptyset$.
- $S_1 \cap S_2$ means that for all tournaments T, $S_1(T) \cap S_2(T) \neq \emptyset$. (Remark that in this case, $S_1(T) \cap S_2(T)$ defines another tournament solution.)

Many properties of tournament solutions are mentioned in the literature, here are the main ones :

2.3. Monotonicity, Strong Superset Property and Independence of Losers

Definition 2.3.1. : A solution S is *monotonous* if for any tournament $T \in \mathcal{T}(X)$, for any $x \in S(T)$ and for any tournament $T' \in \mathcal{T}(X)$ such that :

$$\begin{cases} T'/X-\{x\} = T/X-\{x\} \\ \forall y \in X, xTy \Rightarrow xT'y \end{cases}$$

one has : $x \in S(T')$.

In other words, a solution is monotonous if, whenever a winner is reinforced, it does not become a loser.

Definition 2.3.2. : A solution S is said to be *independent of the losers* if for any tournament $T \in \mathcal{T}(X)$ and for any tournament $T' \in \mathcal{T}(X)$ such that :

$$\forall x \in S(T),\ \forall y \in X,\ xTy \Leftrightarrow xT'y$$

one has : $S(T) = S(T')$.

In other words, a solution is independent of the losers if the only relations that matter are the relations from winners to winners and the relations from winners to losers. What happens between two losers does not matter.

Notation : Let $T \in \mathcal{T}(X)$, $(x, y) \in X^2$; we denote by $T\langle x, y\rangle$ the tournament on X obtained by reversing the arrow between x and y ; that is, for all u and v in X :

$$\begin{cases} u \neq x \text{ and } u \neq y \Rightarrow [uT_{\langle x,y\rangle}v \Leftrightarrow uTv] \\ u \neq x \Rightarrow [uT_{\langle x,y\rangle}y \Leftrightarrow uTy] \\ u \neq y \Rightarrow [uT_{\langle x,y\rangle}x \Leftrightarrow uTx] \\ xT_{\langle x,y\rangle}y \Leftrightarrow yTx \end{cases}$$

All the considered tournaments being finite, with this notation, the Monotonicity property can be written:

$$\forall (x, y) \in X^2,\ x \in S(T) \text{ and } yTx \Rightarrow x \in S(T\langle x, y\rangle)$$

and the Independence of the losers property can be written :

$$\forall (x, y) \in (X - S(T))^2,\ S(T\langle x, y\rangle) = S(T).$$

<u>Definition 2.3.3.</u> : A solution S satisfies the *Strong Superset Property* (SSP) if for any tournament $T \in \mathcal{T}(X)$ and for any Y such that $S(T) \subset Y \subset X$ one has $S(T) = S(T / Y)$.

In other words, a solution satisfies the Strong Superset Property if one does not change the set of winners by deleting some or all of the losers (Bordes, 1979). These three properties are not logically independent the ones from the others, as stated in the next proposition.

Proposition 2.3.4. : If S is monotonous and verifies *SSP*, then S is independent of the losers.

Proof :

Let $T \in \mathcal{T}(X)$, $(x, y) \in [X - S(T)]^2$, with yTx. By Monotonicity, $y \in S(T\langle x, y\rangle) \Rightarrow y \in S(T)$, thus $y \notin S(T\langle x, y\rangle)$, but obviously $T\langle x, y\rangle \,/\, (X - \{y\}) = T \,/\, (X - \{y\})$, thus by SSP, $S(T\langle x, y\rangle) = S(T)$. ■

Proposition 2.3.5. : The Top-Cycle solution, *TC*

(i) is monotonous

(ii) is independent of the losers

(iii) satisfies SSP.

Proof :

Each of the three points is immediately deduced from the definition of the Top-Cycle. ■

One may think of other requirements concerning monotonicity, leading to alternative definitions. Here is an example. Let $x \in S(T)$ be strengthened : yTx and $T' = T\langle x, y\rangle$. If S is supposed to select one alternative, the fact that $S(T)$ contains several elements is usually interpreted as "according to concept S, elements of $S(T)$ are tied." Then a desirable property is that strengthening x breaks these ties and makes x the sole winner : $\{x\} = S(T')$. Two variants of that "tie-breaking monotonicity" may be used: in the strong variant, one requires $\{x\} = S(T\langle x, y\rangle)$ for any y such that yTx, and in the weak variant, one only requires $\{x\} = S(T\langle x, y\rangle)$ for any winner $y \in S(T)$ such that yTx. Other notions of monotonicity can be studied, replacing "strengthening a winner" by "weakening a loser". For the sake of simplicity, we formally defined only one notion of monotonicity (2.3.1.) and other tie-breaking, monotonicity-like properties will be eventually given when studying the various solutions.

We shall also use two properties, Idempotence and the Aïzerman property, which are two weakenings of the Strong Superset Property.

Definition 2.3.6. : A solution S is *idempotent* if $S \circ S = S$.

Definition 2.3.7. : A solution S satisfies the *Aïzerman Property* if for any tournament $T \in \mathcal{T}(X)$ and any Y:

$$S(T) \subset Y \subset X \text{ implies } S(T/Y) \subset S(T).$$

The tournament $T/(X\text{-}\{x\})$ obtained by deleting the alternative x will be denoted by $T\text{-}x$. Because we only deal with finite tournament, the Aïzerman property can be stated : if x is an S-loser, $S(T\text{-}x) \subset S(T)$. The two following propositions about the Aïzerman property will be useful.

Proposition 2.3.8. : Let S_1 and S_2 be two solutions such that S_1 satisfies the Aïzerman property and is more selective than S_2, then for any $k>0$, including $k=\infty$, S_1^k is more selective than S_2^k.

Proof :

Let $T \in \mathcal{T}(X)$. In order to simplify notations, we write, the tournament T being fixed, $S(Y)$ instead of $S(T/Y)$. Let us prove by induction on k the following inclusions :

$$\begin{cases} S_1^k \subset S_1^{k-1} o\, S_2 \subset S_1^{k-1} \\ S_1^k \subset S_2^k \end{cases}$$

For k = 1, they are true by hypothesis. Suppose they are true for $k \geq 1$, then the Aïzerman property applied to $S_1^{k-1}(T)$ and $S_1^{k-1} o\, S_2(T)$ gives:

$$S_1^k\, o\, S_2 \subset S_1^k \text{, thus : } S_1^k\, o\, S_2 \subset S_1^k \subset S_1^{k-1}\, o\, S_2.$$

The Aïzerman property applied to $S_1^{k-1} o\, S_2(T)$ and $S_1^k(T)$ then gives:

$S_1^{k+1} \subset S_1^k o\, S_2$, thus: $S_1^{k+1} \subset S_1^k o\, S_2 \subset S_1^k$.

Moreover $S_1^k \subset S_2^k$ implies $S_1^k o\, S_2 \subset S_2^{k+1}$, and because $S_1^{k+1} \subset S_1^k o\, S_2$, we have $S_1^{k+1} \subset S_2^{k+1}$, which completes the induction. The inclusion $S_1^k \subset S_2^k$ being true for any finite k, one easily proves $S_1^\infty \subset S_2^\infty$, this concludes the proof of the proposition. ■

<u>Proposition 2.3.9.</u> : Let S be a solution which satisfies the Aïzerman property, let $T \in \mathcal{T}(X)$ and $x \in X$, with $x \notin S(T)$, then $S^\infty(T/X - \{x\}) = S^\infty(T)$.

Proof :

Let $T' = T/X - \{x\}$, we prove by induction on k that:

$S^k(T) \subset S^{k-1}(T') \subset S^{k-1}(T)$.

For $k = 1$, this reads : $S(T) \subset X - \{x\} \subset X$, which is true. Suppose that the inclusions are true for $k \geq 1$, then the Aïzerman property gives :

$S^k(T') \subset S^k(T) \subset S^{k-1}(T')$.

Using a second time this same property then gives:

$S^{k+1}(T) \subset S^k(T') \subset S^k(T)$,

which concludes the induction. The result follows. ■

The proposition 2.3.9. allows for an *algorithmic* determination of the solution set $S^\infty(T)$, in the case where S satisfies the Aïzerman property. The algorithm is simply : "Find a S-loser and remove it from the set of vertices. Continue until there is no S-loser".

It can be remarked that, unlike monotonicity and independence of the losers, the three properties SSP, Idempotence and Aïzerman do not need in their definitions any reference to the tournament structure. They are indeed properties of abstract choice correspondences and have been introduced in the context of the general Theory of Choice. They are related by the following proposition.

<u>Proposition 2.3.10.</u> : SSP is equivalent to the conjunction of Idempotence and Aïzerman.

Proof :

It is clear that SSP implies the two others properties. On the other hand, let $S(X) \subset Y \subset X$. Using twice the Aïzerman property, one gets : $S(Y) \subset S(X) \subset Y$ and $S^2(X) \subset S(Y) \subset S(X)$. By idempotence it follows : $S(X) = S^2(X) = S(Y)$. ■

Many other properties have been introduced in the literature on the rationalizability of choice (see in particular Sen 1971). The Expansion property (Sen's property γ) states that for any X and Y, $S(X) \cap S(Y) \subset S(X \cup Y)$. It is immediate that this property is satisfied by the Top-Cycle, and we will come back to the Expansion property in the chapter devoted to the Uncovered set (proposition 5.1.6.). The two folowing properties are often mentionned.

Chernoff (or Sen's α) : $Y \subset X \Rightarrow S(X) \cap Y \subset S(Y)$.

Path Independence : $S(X \cup Y) = S(S(X) \cup Y)$.

By considering a three-cycle, the reader will easily check that these properties cannot be satisfied by tournament solutions. The Top-Cycle also satisfies the following property.

Sen's β : $(Y \subset X$ and $S(X) \cap S(Y) \neq \varnothing) \Rightarrow S(Y) \subset S(X)$.

The other considered tournament solutions do not satisfy Sen's β. To see that this property is not really interesting for a tournament solution, observe that if S satisfies Sen's β then if x is a S-winner and x belongs to a three-cycle (x, y, z), y and z must be S-winners too. Therefore, to verify Sen's β, a tournament solution must *not* break the Condorcet cycles.

2.4. Composition-Consistency and Regularity

Definition 2.4.1. : A solution S is *composition-consistent* if for any tournament T and for any decomposition $T = \Pi(T^*; T_1, ..., T_k)$ of T, one has :

$$S(T) = \bigcup_{i\,:\,T_i \in S(T^*)} S(T_i).$$

The meaning of this property is the following : suppose that the alternatives to be decided upon by the society can be classified in projects $X_1, ..., X_k$, each project X_i being possible under several variants $x_{i,\alpha}$, for $\alpha = 1, ..., n_i$. Suppose that if a project X_i is judged better than a project X_j then all the variants of X_i are judged better than all the variants of X_j ; that is to say that the comparison of any two projects is independent of the variants that represent these projects. The composition-consistency property is the requirement of choosing the best variants of the best projects (Laffond, Lainé and Laslier 1996). A similar notion was introduced by Tindeman (1987) under the name "independence of clones" (see also Tindeman and Zavist 1989).

Proposition 2.4.2. : If S_1 and S_2 are two composition-consistent solutions then $S_1 \circ S_2$ is composition-consistent.

Proof :

The tournament restricted to $S_2(T)$ admits a decomposition :

$S_2(T) = \Pi(S_2(T^*) ; ..., S_2(T_i), ...)$, with $T_i \in S_2(T^*)$, thus :

$S_1 \circ S_2(T) = \bigcup_{i\,:\,T_i \in S_1(S_2(T^*))} S_1(S_2(T_i))$. ■

Corollary 2.4.3. : If S is composition-consistent, S^n and S^∞ are composition-consistent too.

<u>Proposition 2.4.4.</u> : If S is composition-consistent, $S \subset TC$.

Proof :

In effect, if T is reducible, according to the proposition 1.3.13., $TC(T)$ is the Condorcet winner of a summary of T. ■

Several weakenings of the composition-consistency property can be defined. We shall consider the following one, used by Moulin (1986). Let Y be a component of T. For composition-consistency, one envisages replacing T/Y by any tournament. But one may also envisage changing the tournament T/Y without changing the set of alternative Y. The difference is that arrows can be reversed in T/Y, but the order of the component is not allowed to vary. Specifically, we introduce the following definition.

<u>Definition 2.4.5.</u> : A solution S is *weakly composition-consistent* if for any $T \in \mathcal{T}(X)$, if Y is a component of T and $x, y \in Y$:

$$S(T) \cap (X\text{-}Y) = S\,(T\langle x, y\rangle) \cap (X - Y) \text{ and}$$

$$S(T) \cap Y \neq \varnothing \Rightarrow S\,(T\langle x, y\rangle) \cap Y \neq \varnothing.$$

Clearly, if S is composition-consistent, then S is weakly composition-consistent. The next property is not a weakening of composition-consistency ; remember that a regular tournament is a tournament in which all the alternatives beat (and are beaten by) the same number of alternatives.

<u>Definition 2.4.6.</u> : One says that a solution S is *regular* if for all regular tournaments $T \in \mathcal{T}(X)$, $S(T) = X$.

Contrary to the intuition, there exist reasonable tournament solutions which are not regular (see chapter 7). This means that there exist tournaments in which each alternative beats exactly half and is beaten exactly by half of the other alternatives, and in which some alternatives may be said "better", in a reasonable sense, than the others.

Proposition 2.4.7. : If a solution S is composition-consistent and regular, then for any tournament $T \in \mathcal{T}(X)$, if T is the summary of some regular tournament, then $S(T) = X$.

Proof :

This is a straightforward application the two definitions of composition-consistency and regularity. ■

Proposition 2.4.8. : The Top-Cycle solution, TC,

(i) is regular

(i) is not composition-consistent.

Proof :

(i) If $TC(T) \neq X$ let $n = \# TC(T)$, $m = \#(X - TC(T))$ and $y \in X\text{-}TC(T)$. Then $s(y) \leq m - 1$, and there exists $x \in TC(T)$ such that $s(x) \geq En[\frac{n-1}{2}] + m > s(y)$ (we denote by $En[]$ the integer part).

(ii) A counter-example is given by the tournament in figure 2.1, 1' is in the Top-Cycle of the tournament but is not in the Top-Cycle of the component {1, 1'}. ■

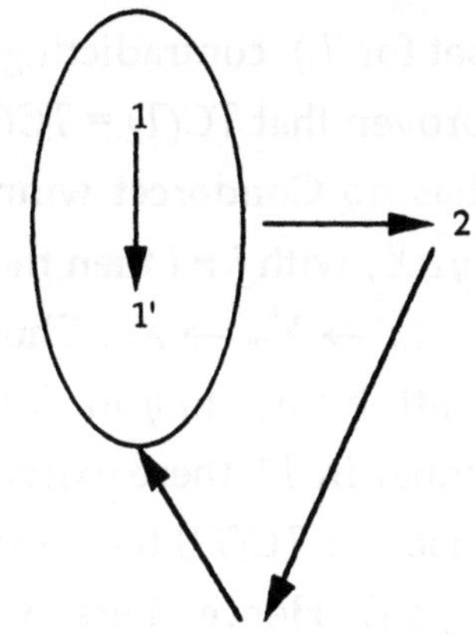

Figure 2.1

Even if *TC* is not composition consistent, it is easy to compute the Top-Cycle of a composed tournament.

Proposition 2.4.9.: Let $T = \Pi(T^*; T_1, ..., T_k)$ be a composed tournament. If T^* has a Condorcet winner, say X_i, then $TC(T) = TC(T_i)$. If T^* has no Condorcet winner, then $TC(T) = \cup\{X_i \in TC(T^*)\}$.

Proof :

Suppose that T^* has a Condorcet winner X_i and let $x \in TC(T_i)$ and $y \in X = \bigcup_{i=1}^{n} X_i$. If $y \in X_i$, then $x \in TC(T_i)$ implies that there exists a path : $x \to x_1 \to ... \to y$ from x to y in T_i. If $y \in X_j$, $j \neq i$ then $X_i\, T^*\, X_j$ implies that xTy. Hence in any case there is a path from x to y, and $x \in TC(T)$. Conversely, let $x \in TC(T)$, consider a shortest path $x = x_0 \to x_1 \to ... \to x_m \in TC(X_i)$ from x to $TC(T_i)$. The length of such a path is $m \geq 0$; if $x \notin TC(T_i)$, that is if $m > 0$, one finds that $x_{m-1} \in X_i$ (because X_i is a Condorcet winner in T^*) and therefore, that $x_{m-1} \in TC(T_i)$ (because

$TC(T_i)$ is a retentive subset for T_i), contradicting the minimality of m. Thus $x \in TC(T_i)$ and it is proven that $TC(T) = TC(T_i)$.

Suppose that T^* has no Condorcet winner and let $x \in X_i$ and $y \in X$, with $X_i \in TC(T^*)$. If $y \in X_j$, with $j \neq i$ then there exists a path in T^* of the form : $X_i \to X_{i1} \to \ldots \to X_{im} \to X_j$. Choosing one element in each $X_{i1}, \ldots, X_{im}$ gives a path from x to y in T. If $y \in X_i$ then, because there is no Condorcet winner in T^*, there exists $X_j \in TC(T^*)$ such that $X_i\ T^*\ X_j$. Take $z \in X_j$, because $X_j \in TC(T^*)$ there exists a path from z to y in T, and because $X_i\ T^*\ X_j$, xTz. Hence, there exists a path from x to y in T. This proves that $x \in TC(T)$. Conversely, if $x \in TC(T)$ and $x \in X_i$, for any $j \neq i$ take $y \in X_j$ and consider a path from x to y in T, the components to which the elements of this path belong form a path from X_i to X in T^*, therefore, $X_i \in TC(T^*)$. This completes the proof of the proposition. ■

If Y is a component of T, one can consider the decomposition $\{Y\} \cup \{\{x\}: x \in X-Y\}$ in which all components, apart from Y, are singletons. By applying the preceding proposition, one obtains for the Top-Cycle :

<u>Corollary 2.4.10.:</u> The solution TC is weakly composition-consistent.

2.5. Composition-Consistent Hulls

The composition-consistency property describes the sensitivity of a solution to cloning and/or contraction of the set of alternatives. In order to study more precisely this sensitivity, we introduce the concepts of composition-expansion and composition-consistent hull of a tournament solution. A given alternative x belongs to the composition-expansion of a tournament solution S if it is possible to clone the alternatives in such a way that some point of the now-component x is chosen.

__Definition 2.5.1.:__ Let T be a tournament defined, for the simplicity of notations, on $\{1, ..., K\}$. Let S be a tournament solution. The *composition-expansion* of S, denoted $\widetilde{S}$ is the tournament solution defined by : $i \in \widetilde{S}(T)$ if there exist K tournaments $T_1, ..., T_k$ on K disjoints sets $X_1, ..., X_k$ such that $S(\widetilde{T}) \cap X_i \neq \varnothing$, where $\widetilde{T} = \Pi(T; T_1, ..., T_k)$.

Clearly $\widetilde{S}$ is a tournament solution and $S \subset \widetilde{S}$. But $\widetilde{S}$ is not always composition-consistent. For instance, it is easy to verify that for the Top-Cycle, $\widetilde{TC} = TC$. So the composition-extension of S does not capture the idea "What must one choose if one wants S-winners to remain winners and the composition-consistency property to hold simultaneously?" We need the following definition, introduced by a useful proposition.

__Proposition 2.5.2.:__ Let $(S_a)_{a \in A}$ be a (possibly infinite) family of composition-consistent tournament solutions. If, for any tournament T, $S'(T) = \bigcap_{a \in A} S_a(T) \neq \varnothing$ then S' is a composition-consistent tournament solution.

Proof :

Let $T = \Pi\ (T^*; T_1, ..., T_k)$ for $x \in X_i$, $x \in S'(T)$ iff for all $a \in A$, $i \in S_a(T^*)$ and $x \in S_a(T_i)$ hence $S'(T) = \bigcup_{i \in S'(T^*)} S'(T_i)$ and the result. ■

Thanks to this proposition, one can speak of the finest composition-consistent tournament solution that contains a given tournament solution S, it is the intersection of all the composition-consistent tournament solutions coarser than S.

Definition 2.5.3.: The *composition-consistent hull* of a tournament solution S is the tournament solution S^* such that $S \subset S^*$ and, for any composition-consistent solution S', $S \subset S' \Rightarrow S^* \subset S'$.

The composition consistent hull is related to the composition-extension by the following proposition.

Proposition 2.5.4.: Let S be a tournament solution, $\widetilde{S}$ its composition-expansion and S^* its composition-consistent hull. Then $\widetilde{S} \subset S^*$ and therefore, $\widetilde{S} = S^*$ if $\widetilde{S}$ is composition-consistent.

Proof :

Let $i \in \widetilde{S}(T)$, $\widetilde{T} = \Pi(T; T_1, ..., T_k)$ such that $S(\widetilde{T}) \cap X_i \neq \emptyset$. Let $x \in S(\widetilde{T}) \cap X_i$ and consider S' a composition-consistent solution such that $S \subset S'$: $x \in S(\widetilde{T}) \cap X_i$ and $i \in S'(T)$. Thus $i \in S^*(T)$ and we just proved that $\widetilde{S}(T) \subset S^*(T)$. Then, by definition of S^*, $S^* = \widetilde{S}$ if $\widetilde{S}$ is composition-consistent. ■

Proposition 2.5.2.: The composition-consistent hull of the Top-Cycle, TC^* is the degenerated tournament solution such that $\forall T \in \mathcal{T}(X)$, $TC^*(T) = X$.

Proof :

Let $T \in \mathcal{T}(X)$, $a \notin X$, $b \notin X$; consider $\widetilde{T} = \Pi(C_3; T, T_a, T_b)$ where C_3 is the cyclone of order 3, T_a and T_b are the two trivial tournaments on $\{a\}$ and $\{b\}$. Then $TC(\widetilde{T}) = X \cup \{a\} \cup \{b\}$ hence $TC^*(T) = X \cup \{a\} \cup \{b\}$. This implies by composition-consistency that $TC^*(T) = X$. ■

3 - Scoring and Ranking Methods

In order to determine the winners of a tournament, a first possibility is to associate to each alternative a number reflecting the quality of this alternative, this gives a ranking, then to choose the best alternatives according to this ranking. The most natural method is the so-called Copeland score, where one simply counts the number of wins of each alternative. But other methods have been proposed (see Jech 1983, 1989). It is impossible in one chapter to study them all in full details, but most can be considered as variants of one of the three scoring methods we study: the Copeland, Long-Path and Markov scores. The chapter also discusses the Slater method for ranking the alternatives ; this widely held method directly defines optimal rankings without reference to a notion of score.

3.1. Copeland Solution

Recall that the Copeland score of an alternative x in T is the number of alternatives beaten by x : $s(x) = \#T^{+}(x)$. A Copeland winner of T is an alternative with highest Copeland score.

<u>Definition 3.1.1.</u>: The set of *Copeland winners* of $T \in \mathcal{T}(X)$ is :

$$C(T) = \{x \in X : \forall y \in X, s(y) \leq s(x)\}.$$

The set of Copeland winners clearly defines a tournament solution. The properties of this solution are given now.

<u>Theorem 3.1.2.</u>: The Copeland solution, C

(i)	is monotonous
(ii)	is not independent of losers
(iii)	does not verify SSP
(iv)	is not idempotent
(v)	is not composition-consistent
(vi)	is not weakly composition-consistent
(vii)	is regular
(viii)	is included in the Top-cycle.

Proof :

The points (i), (vii) and (viii) are trivial, the point (iii) is deduced from the point (iv) and (v) is deduced from (vi), so remain (ii), (iv) and (v). For this, consider again the tournament T in figure 2.1. Let T' be obtained from T by changing the arrow $1' \rightarrow 2$ to $2 \rightarrow 1'$:

$$T = \{(1, 1'), (1, 2), (1', 2), (2, 3), (3, 1), (3, 1')\}$$
$$T' = \{(1, 1'), (1, 2), (2, 1'), (2, 3), (3, 1), (3, 1')\}$$

Then $C(T) = \{1, 3\}$ and $C(T') = \{1, 2, 3\}$, which proves (ii). By deleting the vertices $1'$ and 2 of the tournament T one gets :

$T'' = \{(3, 1)\}$, $C(T'') = \{3\}$, which proves (iv). In order to prove (vi), one takes three points $(1, 1', 1'')$ instead of two in the component, $X = [1, 1', 1'', 2, 3\}$ and $\{\{1, 1', 1''\}, \{2\}, \{3\}\}$ a decomposition, $\{1, 1', 1''\} \rightarrow \{2\}$, $\{2\} \rightarrow \{3\}$ and $\{3\} \rightarrow \{1, 1', 1''\}$. Then if the component $\{1, 1', 1''\}$ is cyclic the unique Copeland winner is *3* and if it is transitive, say $1 \rightarrow 1' \rightarrow 1''$, $1 \rightarrow 1''$ there are two Copeland winners: *1* and *3*. ■

An axiomatization of the ranking of the alternatives by mean of their Copeland score is given by Rubinstein (1980). Let $R(T) \in \mathcal{OC}(X)$ be a complete ordering of X associated to tournament $T \in \mathcal{T}(X)$ and let $P(T)$ be the asymmetric part of $R(T)$. Consider the three following requirements:

a) For all permutations $\varphi \in \mathcal{O}(X)$ of X and all x, y in X, $xR(T)y$ if $\varphi(x)\ R(\varphi^T)\ \varphi(y)$, where φ^T is the tournament on X defined by $\varphi(x)\ (\varphi^T)\ \varphi(y)$ iff xTy.

b) Let x and y be distinct alternatives in X such that $xR(T)y$, let $z \in X$ be such that zTx and let $T' = T_{<x,z>}$ be the tournament on X identical to T except that $xT'z$. Then $xP(T')y$.

c) Let x, y, u and v be four distinct alternatives in X, then $uR(T)v$ if $uR(T_{<x,y>})v$.

Then R satisfies a), b) and c) if and only if R corresponds to the ranking by the Copeland scores: in the tournament T, $xR(T)y$ iff $s(x) \geq s(y)$.

Statement a) is a neutrality statement. Statement b) is a strong monotonicity requirement, it implies that the Copeland tournament solution satisfies that if a Copeland winner is strengthened then it becomes the unique Copeland winner. As will be seen in the sequel, other proposed scoring methods satisfy b), whereas requirement c) is controversial and is not met by standard scoring methods -- apart from the Copeland method. Another axiomatization of the Copeland rule is given by Henriet (1985).

The ranking of the alternatives according to the Copeland score is also identical to the ranking obtained by an apparently different method, proposed by Zermelo (1929) and called the maximum likelihood method. Suppose that to each alternative $x \in X$ is attached a weight $v(x)$ such that, when comparing x and y, the probability that x beats y is equal to $\frac{v(x)}{v(x)+v(y)}$. Supposing the results of the various meetings independent the ones from the others, the probability of observing the tournament T is equal to :

$$\prod_{xTy} \frac{v(x)}{v(x)+v(y)} = \frac{\prod_{x} [v(x)]^{s(x)}}{\prod_{\{x,y\}} (v(x)+v(y))}$$

If T is irreducible (only interesting case) there is a unique vector v which maximizes this probability. Those weights are not proportional to the Copeland scores, but rank X in the same way (see Moon 1968). The models of Bradley-Terry (1952) and Ford (1957) are variants of Zermelo's idea.

3.2. Iterative Matrix Solutions

Let T be a tournament of order n defined (without loss of generality) on $\{1, \ldots, n\}$. The *matrix of* T is the square matrix of order n, $M = (M_{xy})$, with $M_{xy} = 1$ if xTy and $M_{xy} = 0$ if not. (In particular, the diagonal of M is null). The study of tournaments can be performed using such matrices $M \in \{0, 1\}^{n^2}$ with $M_{xx} = 0$ and $M_{xy} = 1 \Rightarrow M_{yx} = 0$. If 1 denotes the column vector of 1, the vector of the Copeland scores is $s = M1$. Several propositions have been made for defining more sophisticated scores. One can for instance be interested by the behavior of $M^t 1$ when t tends to infinity, or better $(M + 1/2\ I)^t 1$, where I is the 1-diagonal matrix, which means counting 1/2 instead of 0 for the comparison of x with itself. Those variants can give rankings which are different from the Copeland one. The original references given by Moon are Wei (1952) Kendall (1955), Katz (1953), Thompson (1958) ; on tournament matrices see Moon and Pullman (1967) and Maybee and Pullman (1990).

The Long Path method.

Consider for instance the following method, which we shall call the "Long Path Method". We start with the computation of the Copeland scores $s_1=M1$. Then we observe that we have given to an alternative x one point for each win $x \to y$, whereas it could be more fair to give more points to x when x beats a "good" alternative than a "bad" one. So we decide to give to x precisely $s_1(y)$ points for x dominating y and we compute :

$$s_2(x) = \sum_{y: x \to y} s_1(y)$$

that is: $s_2=Ms_1=M^2 1$. We can go on and define inductively s_t for $t \geq 1$ by $s_0=1$ and $s_{t+1}=Ms_t=M^{t+1}1$. The entry $M^t{}_{x,y}$ of the matrix M^t is, in the graph of the tournament, the number of paths of length t going from x to y , and $s^t(x)$ is the number of paths of length t starting at x. The Long Path Method consists in ranking the alternatives according to s_t, for t large. This is the method advocated by C. Berge (1970).

If x belongs to a cycle $x=x_1 \to x_2 \to \ldots \to x_k=x$ or if x (indirectly) beats a cycle then $s_t(x)$ never equals zero. In the opposite case, that is when the restriction of T to x and its successors is transitive, $s_t(x)$ is zero for t large enough. So we suppose that the tournament is not transitive and we first look at the case of irreducible tournaments. In particular, for all x and t, $s_{t+1}(x) \geq s_t(x)$ and for no x and no t, $s_t(x)$ is null.

A tournament matrix is non negative, so we can use the Perron-Frobenius theorem in order to study its powers. The matrix M has a unique real eigenvalue λ of maximal absolute value, λ is called the Frobenius root of M. Let lp be an eigenvector associated with λ, $lp \neq 0$ and we can take $\sum_{x \in X} lp(x) = 1^T.lp = 1$ (we denote by 1^T the transposed vector of 1). Since the Frobenius root is simple, $lp(x)$ equals the limit of $s_t(x)/\ 1^T.s_t$, which means that $lp(x)$ can be interpreted as the proportion of "long" paths of the graph of the tournament starting at x. In order to compute lp, one just has to find an eigenvector of M associated to its Frobenius root. The sequence $s_t(x)$, $t \geq 0$ goes to infinity exponentially like λ^t. (Keener (1993) gives various applications of the Perron-Frobenius theorem to the question of ranking football teams.)

Definition 3.2.1.: The Long-Path score $lp(x)$ of $x \in X$ in $T \in \mathcal{T}(X)$ is the entry x of the normalized eigenvector associated to the Frobenius root of the matrix of T.

The set of alternatives with maximal Long-Path score defines a tournament solution, the interested reader will easily (?) study the properties of this solution.

3.3. Markov Solution

Following the same kinds of ideas, we propose another solution, which we shall call the "Markov method". Suppose that at each date $t = 1, \ldots$ there is a match between two alternatives $x(t)$ and $y(t)$, such that $x(1)$ and $y(1)$ are arbitrary and different, and for $t>1$, $x(t)$ is the winner of the match in $t - 1$, and $y(t)$ is chosen at random, uniformly in X-$\{x(t)\}$. One thus defines a finite Markov chain on the space state

X. If $p(t)$ is the column vector of the probabilities for the random variable $x(t)$ to take the values $x \in X$, one has :

$$p(t+1) = \frac{1}{n-1}(M+S)p(t)$$

where S is the diagonal matrix of the Copeland scores. Denote by N this matrix and call it the *transition matrix* of T (each column sums to 1). The elementary theory of finite Markov chains shows that 1 is an eigenvalue of multiplicity 1 of N, that $p(t)$ tends to a positive limit $\overline{p}$ which does not depend on $p(0)$ and which verifies :

$$\overline{p}_x > 0 \Leftrightarrow x \in TC(T).$$

(In the language of the Markov chains, the Top-Cycle is the unique absorbing class of N). Thus the computation of these "Markov scores" is performed by looking at the eigenvector of N associated to the eigenvalue 1 and such that $\sum_{x \in X} \overline{p}_x = 1$; that is to say $\overline{p}$ verifies :

$$\overline{p}_x = \frac{1}{n-1} s^+(x)\overline{p}_x + \frac{1}{n-1} \sum_{y \in T^+(x)} \overline{p}_y ,$$

which can be also written : $s^-(x)\ \overline{p}_x = \sum_{y \in T^+(x)} \overline{p}_y$.

(We write here $s^+(x) = s(x)$ and $s^-(x) = n - 1 - s(x)$).

Definition 3.3.1.: The vector of the *Markov scores*, $\overline{p}$, is the eigenvector of the transition matrix associated to the eigenvalue 1 and such that $\sum_{x \in X} \overline{p}_x = 1$.

A *Markov winner* is a point with highest Markov score.

Denote by $Ma(T)$ the set of Markov winners for the tournament T. If there is a Condorcet winner for T, clearly this Condorcet winner is the Markov winner, with a score of 1, all the other points getting 0. If there is no Condorcet winner then for any $x \in X$:

$$\overline{p}_x \leq s^-(x)\ \overline{p}_x = \sum_{y \in T^+(x)} \overline{p}_y < 1 - \overline{p}_x$$

and $\overline{p}_x < ½$. It can also be observed that a Markov winner beats at least half of the other outcomes: Let $x \in Ma(T)$, then: $s^-(x)\ \overline{p}_x = \sum_{y \in T^+(x)} \overline{p}_y \leq s^+(x)\overline{p}_x$ hence $s^+(x) \geq s^-(x)$. This solution is really different from the Copeland one (and also different from the other matrix solutions), as will be seen in the following example :

Let T_{10} be the tournament of order 10 depicted in figure 3.1, where C_3 and C_5 are cyclones of order 3 and 5 :

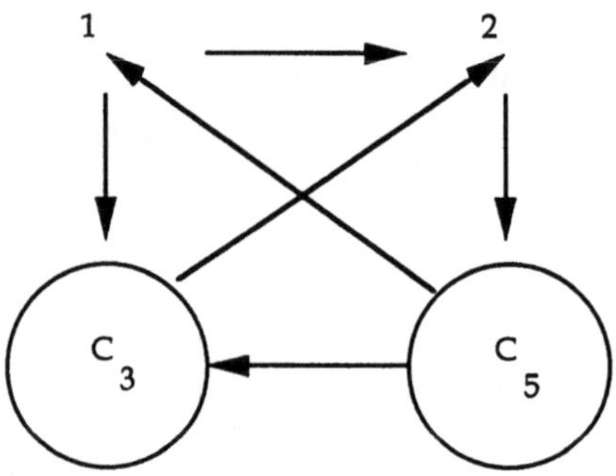

Figure 3.1

Let i be an element of the component C_3 and j be an element of the component C_5. The Copeland scores are given by

$s(1) = 4$, $s(2) = 5$, $s(i) = 2$ and $s(j) = 6$.

Thus one has : $C(T_{10}) = C_5$. Let us compute the Markov scores; one has :

$$\begin{cases} 5\overline{p}_1 = 3\overline{p}_i + \overline{p}_2 \\ 4\overline{p}_2 = 5\overline{p}_j \\ 7\overline{p}_i = \overline{p}_i + \overline{p}_2 \\ 3\overline{p}_j = \overline{p}_1 + 3\overline{p}_i + 2\overline{p}_j \end{cases}$$

which writes : $\begin{cases} 5\overline{p}_1 = 3\overline{p}_i + \overline{p}_2 \\ 4\overline{p}_2 = 5\overline{p}_j \\ 6\overline{p}_i = \overline{p}_2 \\ \overline{p}_j = \overline{p}_1 + 3\overline{p}_i \end{cases}$

The solution is $\overline{p}_1$=9/174, $\overline{p}_2$=30/174, $\overline{p}_i$=5/174 and $\overline{p}_j$=24/174. As a consequence, *2* is the unique Markov winner of T_{10}. This demonstrates that the Copeland and Markov solutions may have an empty intersection :

Proposition 3.3.2.: $C \varnothing Ma$

Following Levchenkov (1992), let us study the monotonicity properties of the Markov solution. Let *a*, $b \in X$ with *aTb*. We wish to know how the values of $\overline{p}_a$, $\overline{p}_b$, and $\overline{p}_x$, $x \neq a, b$ are affected when the arrow $a \rightarrow b$ is reversed. The only interesting case is when both tournaments *T* and $T_{<a,\,b>}$ are irreducible. The equation $\overline{p} = N\overline{p}$ can be written $B\,\overline{p} = 0$ with $B = \mathbf{I}\text{-}N = \mathbf{I} - \frac{1}{n-1}(M+S)$ that is: $B_{xy} = \frac{-1}{n-1}$ if xTy, $B_{xy} = 0$ if yTx and $B_{xx} = 1 - \frac{s^+(x)}{n-1} = \frac{s^-(x)}{n-1}$. Denote $X' = X\text{-}\{a, b\}$. Line *x* of equation $B\,\overline{p} = 0$ writes:

$$\forall x \in X', \ \sum_{y \in X'} B_{xy}\overline{p}_y = -B_{xa}\overline{p}_a - B_{xb}\overline{p}_b \qquad (1)$$

Call B' the $(n-2) \times (n-2)$ matrix obtained by deleting rows and columns *a* and *b* in *B*. One has the following.

Lemma 3.3.3.: If *T* is irreducible then B' is invertible.

Proof :

To prove this lemma, suppose that B' is not invertible. There exists $q \neq 0$ a (row) vector such that $qB' = 0$. One can suppose that

$\underset{z\in X'}{Max}\, q_z > 0$. Let $y \in X'$ such that $q_y = \underset{z\in X'}{Max}\, q_z$. For any $x \neq y$, $B_{xy} \leq 0$, thus: $-q_x B_{xy} \leq -q_y B_{xy}$, with equality only if $q_x = q_y$ or if $B_{xy} = 0$. Let us sum up these n-3 inequalities (for $x \in X' - \{y\}$).

Because $qB' = 0$, $\sum_{x\in X'-\{y\}} -q_x B_{xy} = q_y B_{yy}$. On the other hand, because $\sum_{x\in X} B_{xy} = 0$ one gets $\sum_{x\in X'-\{y\}} (-B_{xy}) = B_{yy} - B_{ay} - B_{by}$, hence $\sum_{x\in X'} (-B_{xy}) \leq B_{yy}$, with equality only if $B_{ay} = B_{by} = 0$. As a consequence, one may write:

$$q_y B_{yy} = \sum_{x\in X'-\{y\}} (-q_x B_{xy}) \leq q_y \sum_{x\in X'-\{y\}} (-B_{xy}) \leq q_y B_{yy}$$

with equality only if $B_{ay} = B_{by} = 0$ and for all $x \in X' - \{y\}$, $q_x = q_y$ or $B_{xy} = 0$.

Let $Y = \{y \in X' : q_y = \underset{z\in X'}{Max}\, q_z\}$, one must have for any $y \in Y$: yTa (because $B_{ay} = 0$), yTb, and yTx for any $x \in X' - Y$. Thus T is reducible and the lemma is proved. ■

Let C denote the inverse of B', one can write equation (1) for $x \in X'$ in the form:

$$\forall y \in X',\ \bar{p}_y = \sum_{x\in X'} C_{yx} (B_{xa} \bar{p}_a - B_{xb} \bar{p}_b)$$

Denote $\gamma_y^a = -\sum_{x\in X'} C_{yx} B_{xa}$ and $\gamma_y^b = -\sum_{x\in X'} C_{yx} B_{xb}$ then:

$$\forall y \in X',\ \bar{p}_y = \gamma_y^a \bar{p}_a + \gamma_y^b \bar{p}_b \qquad (2)$$

Line a of $B\bar{p} = 0$ is:

$$B_{aa} \bar{p}_a + B_{ab} \bar{p}_b + \sum_{y\in X'} B_{ay} \bar{p}_y = 0'$$

denote $\Gamma_a^a = -\sum_{y\in X'} B_{ay} \gamma_y^a$ and $\Gamma_a^b = -\sum_{y\in X'} B_{ay} \gamma_y^b$, then:

$$(B_{aa} - \Gamma_a^a) \bar{p}_a + (B_{ab} - \Gamma_a^b) \bar{p}_b = 0 \qquad (3)$$

The condition $\sum_{x\in X} \bar{p}_x = 1$ gives, using (2):

$$(1+\sum_{y\in X'}\gamma_y^s)\bar{p}_a+(1+\sum_{y\in X'}\gamma_y^b)\bar{p}_b=1 \quad (4)$$

Solving (3) and (4), one finds the expressions of $\bar{p}_a$ and $\bar{p}_b$:

$$\bar{p}_a=\frac{1}{(1+\sigma_a)+(1+\sigma_b)\dfrac{B_{aa}-\Gamma^a_a}{B_{ab}-\Gamma^b_a}} \text{ and } \bar{p}_b=\frac{1}{(1+\sigma_b)+(1+\sigma_a)\dfrac{-B_{ab}+\Gamma^b_a}{B_{aa}-\Gamma^a_a}}$$

with $\sigma_a=\sum_{y\in X'}\gamma_y^a$ and $\sigma_a=\sum_{y\in X'}\gamma_y^b$. (5)

Before going further, we need to check the signs of the coefficients in expression (5). The following lemma is useful.

<u>Lemma 3.3.4.</u>: Let B' be invertible and such that for all i, $B'_{ii}\geq 0$ and for all $j\neq i$, $B'_{ij}\geq 0$. Let N be a positive vector and γ be such that $B\gamma=N$. Then γ is positive.

Proof:

To prove this lemma, we iteratively solve the system $B\gamma=N$. The first equation of this system can be written:

$$B_{11}\gamma_1=N_1-\sum_{j>1}B'_{nj}\gamma_j$$

Substituting γ_1 in the other equations gives on line i:

$$\sum_{j>1}A_{ij}\gamma_j=B_{11}N_i-B_{i1}N_1$$

with $A_{ij}=B_{11}B_{ij}-B_{i1}B_{1j}$. For $i\neq j$ $A_{ij}\leq 0$ and for $i=j$, $A_{ii}\geq 0$. Moreover, $B_{11}N_i-B_{i1}N_1\geq 0$. Hence (by induction), $\gamma_i\geq 0$ for all $i>1$. Expression for γ_1 proves that $\gamma_1\geq 0$, hence the lemma. ■

Now, because γ^b is positive, Γ^b_a is positive and $B_{ab}-\Gamma^b_a$ is negative. Then (3) shows that $B_{aa}-\Gamma^a_a$ is positive. Because γ^a and γ^b are positive, so are numbers σ_a and σ_b.

Observe that σ_a, σ_b, Γ^a_a, and Γ^b_a depend only on matrix B', thus they are the same for tournaments T and $T_{<a,\ b>}$. Switching from T to $T_{<a,\ b>}$, B_{aa} strictly increases and $-B_{ab}=N_{ab}$ strictly decreases. Hence, expression (5) shows that $\overline{p}_a$ strictly decreases and $\overline{p}_b$ strictly increases.

One also can write the expressions of $\overline{p}_y$ for $y \neq a, b$:

$$\overline{p}_y = \left(\gamma_y^a + \gamma_y^b \frac{B_{aa} - \Gamma_a^a}{-B_{ab} + \Gamma_a^b} \right) \overline{p}_a = \left(\gamma_y^b + \gamma_y^a \frac{-B_{ab} + \Gamma_a^b}{B_{aa} - \Gamma_a^a} \right) \overline{p}_b$$

The expressions show that, for any $y \neq a, b$, $\frac{\overline{p}_y}{\overline{p}_a}$ is increasing and $\frac{\overline{p}_y}{\overline{p}_b}$ is decreasing. As a consequence, if b is a Markov winner for tournament T, then b not only is a Markov winner for $T_{<a, b>}$, but b is the only Markov winner for $T_{<a, b>}$. Like the Copeland solution, the Markov solution is a "tie-breaking rule".

The properties of the Markov solution are given in the following theorem. Monotonicity has already been proven. Other proofs are left to the reader.

<u>Theorem 3.3.5.</u>: The solution *Ma*

(i)	is monotonous
(ii)	is not independent of losers
(iii)	does not verifies SSP
(iv)	is not idempotent
(v)	is not composition-consistent
(vi)	is not weakly composition-consistent
(vii)	is regular
(viii)	is included in the Top-cycle.

Although *Ma* is not composition-consistent, and not even weakly composition-consistent, for a composed tournament one can compute the probability $\bar{q}_i = \sum_{x \in X_i} \bar{p}_x$ of being in a component X_i knowing only the summary and the orders of the components.

<u>Proposition 3.3.6.</u>: Let $T = \Pi(T^*; T_1, \ldots, T_k)$. Denote by $n_i = \#X_i$ the order of T_i, $i = 1, \ldots, k$. Let $\bar{p}$ be the Markov scores for T and $\bar{q}_i = \sum_{x \in X_i} \bar{p}_x$, $i = 1, \ldots, k$, then q is the unique solution to the system of equations:

$$\forall i \in \{1, \ldots, k\}, \left(\sum_{j:jT^*i} n_j \right) q_i = \left(\sum_{j:iT^*j} q_j \right) n_i$$

$$\text{with } \sum_{i=1}^{k} q_i = 1$$

Proof :

For each $x \in X_i$, $s^-(x)\ \bar{p}_x = \sum_{y:xTy} \bar{p}_y$, adding these equations for all $x \in X_i$ leads to the mentioned equation for q_i. Numbers q_i are the stationary probabilities of a Markov process with no more than one absorbing class, hence they are uniquely defined by the system of equations in 3.3.6. ■

The solution proposed here under the name "Markov solution" is called the Ping-Pong winners in Laslier (1992) because it corresponds to the way in which children sometimes play table-tennis contests : *who wins plays again*, and in the long time, the ones who played (and thus won !) more often is the best player. Several authors proposed this procedure, for instance Daniels (1969) and Ushakov (1976) (see Chebotarev and Shamis (1995) for further references). It is called the Self -Consistent Rule by V. S. Levchenkov (1992, 1995a,b) who extended it in two ways :

First, it is possible to consider multirelations instead of tournaments, in a voting problem, this means in particular that it is possible to take into account the sizes of the majorities.

Second, in the case where the tournament is reducible, any point outside the Top-cycle has a zero Markov score. Thus all the points outside the Top-cycle are considered as equivalent ; by using the technique of topological Markov Chains it is possible to distinguish between them.

The idea of a Markov process defined by the choice of the next alternative as an alternative prefered to the current one by a majority of voters is considered by Ferejohn, McKelvey and Packel (1984) in a continuous-space setting.

3.4. Slater Solution

This solution, proposed by Slater (1961), is based on the idea of approximating a tournament by a linear order. One takes the usual distance between graphs and considers the linear orders at minimal distance from the tournament, then the solution is by definition the set of outcomes which are top-element of one of these closest orders.

Definition 3.4.1.: Let T and T' in $\mathcal{T}(X)$; define

$$\Delta(T, T') = (1/2)\#\{(x, y) \in X^2 : xTy \text{ and } yT'x\}.$$

Then Δ is a distance on $\mathcal{T}(X)$, called the "Slater distance".

The fact that Δ is a distance is a standard result, this distance appears in different parts of discrete mathematics, with or without the coefficient $1/2$. It is usually referred to as the Kendall, or Hamming distance, or the distance of the symmetric difference (see Barthélemy, 1979). We use the denomination Slater distance because we shall only use it with reference to the problem of the Slater solution. Recall that $\mathcal{OS}(X)$ denotes the set of the linear (strict) orders on X, a subset of $\mathcal{T}(X)$.

Definition 3.4.2.: Let $T \in \mathcal{T}(X)$, a *Slater order* for T is a linear order $U \in \mathcal{OS}(X)$ such that $\Delta(T, U) = Min\{\Delta(T, V) : V \in \mathcal{OS}(X)\} = \Delta(T, \mathcal{OS}(X))$. The set of *Slater winners* of T, denoted $SL(T)$ is the set of points of X which are Condorcet winner of a Slater order for T.

Minimizing this distance is a particular case of the *median procedure* studied by Barthélemy and Monjardet (1981). It can be pointed out that, if one takes as a reference not the linear orders on X but the set of tournaments having a Condorcet winner, then the minimization of the distance leads to the Copeland solution. The Slater orders for T have the following properties (Some of them are mentionned by Jacquet-Lagrèze, 1969) :

Proposition 3.4.3.: Let U be a Slater order for $T \in \mathcal{T}(X)$, $n = o(T)$. Write $X = \{x_1, ..., x_n\}$, with $x_1 \, U \, x_2 \, U \, ... \, U \, x_n$.

(i) $\Delta(T, U) \leq En[\dfrac{n(n-1)}{4}]$

(ii) $s(x_1) \geq En[n/2]$

(iii) $1 \leq i < j \leq n \Rightarrow U/\{x_i, ..., x_j\}$ is a Slater order for the restriction $T/\{x_i, ..., x_j\}$, and in particular $x_i T x_{i+1}$.

(iv) If Y is a component of T, then U/Y is a Slater order for T/Y.

Proof :

(i) Let $-U$ be the reversed order of U, then clearly, $\Delta(T, U) + \Delta(T, -U) = \dfrac{n(n-1)}{2}$ and $\Delta(T, -U) \geq \Delta(T, U)$, hence (i).

(ii) Let U' be the order obtained, starting from U and letting x_1 go from the first to the last position, then one has : $\Delta(T, U') = \Delta(T, U) + s^+(x_1) - s^-(x_1) = \Delta(T, U) + 2s^+(x_1) - n + 1$, hence $s(x_1) \geq (n-1)/2$. If n is odd, $(n-1)/2 = En[n/2]$. If n is even, then $s(x_1) \geq n/2 = En[n/2]$ because $s(x_1)$ is an integer.

(iii) In effect, re-ordering the segment $\{x_i, ..., x_j\}$, does not change the set of pairs (x, y) such that xTy and yUx with x or y outside $\{x_i, ..., x_j\}$.

(iv) Same argument as for (iii). ■

Proposition 3.4.4.: Let $T \in \mathcal{T}(X)$ be a composed tournament and consider $\{X_1, ..., X_p\}$ a decomposition of T. There exists a Slater order U for T such that each X_i is an interval for U.

Proof :

For $Y \subset X$, let $Max(Y)$ and $Min(Y)$ denote the maximal and minimal elements for U in Y. If yUx, $]x, y[$ denotes the interval $\{z \in X : y\ U\ z\ U\ x\}$ and $[x, y] =]x, y[\cup \{x\ y\}$. Let X_i be a component for T which is not an interval for U, a Slater order for T. Let $x_0 = Max(X_i)$ and $x_1 = Min(\{x \in X_i : [x, x_0] \subset X_i\})$, then $x_1 \neq Min(X_i)$ so there exists an element, say x_2, which follows x_1 in X_i according to U : $x_2 = Max\ \{x \in X_i : x_1\ U\ x\}$. From point (ii) of proposition 3.3.3, U is a Slater order for T restricted to $[x_2, x_1]$. Let $n = \#\ ]x_2, x_1[$ and $m = \#\ \{y \in]x_2, x_1[: y\ T\ X_i\}$. If $m > (n-1)/2$ then one could strictly lower the Slater distance by letting x_1 go down just above x_2, impossible. Likewise, $m < (n-1)/2$ is impossible. Thus $m = (n-1)/2$, and the order obtained from U by letting x_2 go up just underneath x_1 is still a Slater order for T. One can repeat this manipulation until X_i is an interval for U. The other components which already were intervals remain so, and one constructs this way a Slater order in which all the components are intervals. ■

Observe that the prededing proposition does not mean that for a composed tournament, a component is an interval for all the Slater orders. Furthermore, intuition could lead to deduce from this proposition and point (iv) of proposition 3.3.3 that the Slater solution is composition-consistent. In fact this is not the case, as will be stated in the next theorem. In order to find the Slater orders of a composed tournament one has to find the Slater orders of each component and then to order the components taking into account the $\#Y.\#Z$ arrows between components Y and Z.

<u>Theorem 3.1.2.</u>: The Slater solution, *SL*

(i)	is monotonous
(ii)	is not independent of losers
(iii)	does not verify SSP
(iv)	is not idempotent
(v)	is not composition-consistent
(vi)	is weakly composition-consistent
(vii)	is regular
(viii)	is included in the Top-cycle.

Proof :

(i) Let $x \in SL(T)$, $y \in X$ such that yTx. Let U be a Slater order for T with x on top and let U' be a Slater order for $T_{<x, y>}$. Then $\Delta(T_{<x, y>}, U) = \Delta(T, U) - 1$ and moreover $\Delta(T_{<x, y>}, T) = 1$ thus

$$\Delta(T, U) \leq \Delta(T, U') \leq \Delta(T_{<x, y>}, U') + 1,$$

hence $\Delta(T_{<x, y>}, \mathcal{OS}(X)) \geq \Delta(T, U) - 1 = \Delta(T_{<x, y>}, U)$, which proves that U is a Slater order for $T_{<x, y>}$.

(ii), (iii) and (iv) Consider the tournament depicted in figure 3.2.

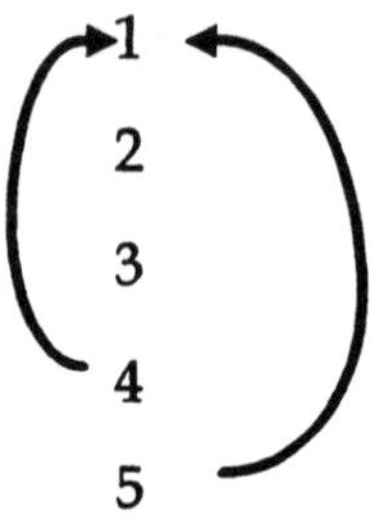

Figure 3.2

It is easy to verify that, for this tournament T, $SL(T) = \{1, 2\}$, and $SL(SL(T)) = \{1\}$, which proves (iii) and (iv). If one reverses the arrow $3 \to 4$, one can verify that for the tournament $T_{<3, 4>}$, a Slater order is $4\ U\ 1\ U\ 2\ U\ 3\ U\ 5$, thus $4 \in SL(T_{<3, 4>})$, which proves point (ii).

(v) A counter-example is given by the tournament depicted in figure 2.1 and already used about the Top-Cycle.

(vi) The proof will be given later (proposition 3.4.8).

(vii) Let $U \in \mathcal{OS}(X)$, x the top-element of U, and U' obtained from U by letting x go from first to last position. If $s(x) = (n-1)/2$ then $\Delta(T, U) = \Delta(T, U')$, hence the result.

(viii) Follows from point (iii) of proposition 3.4.3. Note that the elements of X ranked according to a Slater order form a Hamiltonian path for T. ■

It can be pointed out that, in his seminal paper, Patrick Slater did not aim at defining a ranking procedure. The question he considered was, given a tournament : "Is the hypothesis that this tournament is a randomly perturbed linear order tenable ?" (See also Remage and Thompson 1964.) In order to answer this question, he argues that one has to count the minimal number of arrows that need to be reversed before getting a linear order. So for him the important object is not the order at minimal distance but the minimal distance itself, considered as an indice of (non)-transitivity. He gives arguments against the point of view of Kendall, according to whom a better indice of transitivity is the number of 3-cycles in the tournament. The definitions of the indices of transitivity are not directly relevant for the topic of this book and we refer the reader to the works mentionned in the paragraph on linear tournaments (in chapter 1, section 2) and to Monjardet (1973), Hardouin-Duparc (1975) and Charon-Fournier, Germa and Hudry (1992a, b). Note that the number of 3-cycles in a tournament can be computed knowing the score vector of the tournament, whereas it is difficult to find the orders at minimum distance from a tournament as well as the distance itself.

Several methods have been proposed for determining Slater orders. Remage and Thompson (1966), Bermond and Kondratoff (1976), Barthélémy, Guénoche and Hudry (1989) and Guénoche, Vanderputte-Riboud and Denis (1994) use Bound and Branch algorithms, while Cani (1969) or Marcotorchino and Michaud (1979) use Linear Programming ; see also Junger (1985) or Hudry (1985). Charon-Fournier, Hudry and Woigard (1996) is a survey on the combinatorial and algorithmic aspects of the median orders and Slater orders of the tournaments which contains more references. Bernard Monjardet has written a bibliographical note on the relations at minimal distance from a binary relation (Monjardet 1979).

As a tournament solution, the Slater solution is really different from the other proposed solutions.

<u>Proposition 3.4.6.</u>: (i) $C \varnothing SL$.

(ii) $Ma \varnothing SL$.

Proof :

The first counter-example for (i) is attributed to Bermond (1972). The following example will prove (i) and (ii). Consider the 7-points tournament depicted in figure 3.3. This figure shows an ordering at distance 2 from the tournament because, with the usual convention of overlooking the arrows going down, two arrows had to be drawn.

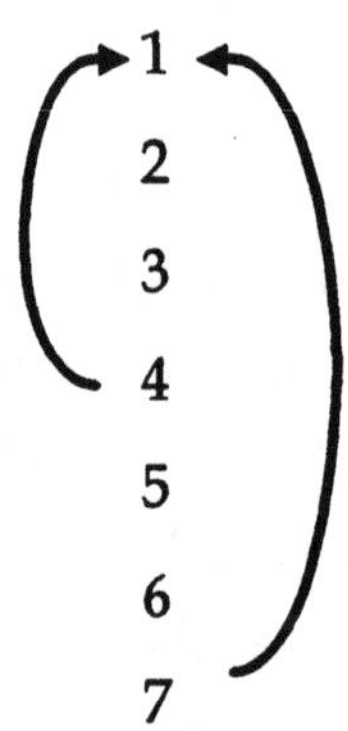

Figure 3.3

Computation of the Slater solution : It is not difficult to check that the order shown on figure 3.3 is the unique Slater order for this tournament. Therefore *1* is the unique Slater winner.

Computation of the Copeland solution : The unique Copeland winner is 2, with a Copeland score of 5.

Computation of the Markov solution : These scores are approximately :

$p(1) = 0.26$
$p(2) = 0.37$
$p(3) = 0.12$
$p(4) = 0.17$
$p(5) = 0.02$
$p(6) = 0.01$
$p(7) = 0.05$

and the unique Markov winner is 2. ■

How to compare the merits of the Slater solution with the ones of the Copeland and Markov solution ? It appears that the Copeland and Markov solutions have very nice monotonicity-like tie-breaking properties which the Slater method does not share. On the other hand, the Slater solution, unnlike the two others, is weakly composition-consistent.

If one strengthens a winner then, according to Copeland and Markov, this alternative is the unique winner of the modified tournament. As the reader will easily verify, this is not always true for the Slater solution. Nevertheless, the following property holds :

<u>Proposition 3.4.7.</u>: Let $x \in SL(T)$, $y \in X$ with yTx, then $SL(T_{<x,\,y>}) \subset SL(T)$ and if $y \in SL(T)$ then $y \notin SL(T_{<x,\,y>})$.

Proof :

Let U be a Slater order order for T, with x on top. Denote $T' = T_{<x,\,y>}$, then $\Delta(T', U) = \Delta(T, U) - 1$ and U is a Slater order for T'. Let $z \in SL(T')$ and U' a Slater order for T' with z on top. If $xU'y$ then $\Delta(T, U') = \Delta(T', U') + 1$ proves that U' is also a Slater order for T and hence $z \in SL(T)$. Suppose now that $y \in SL(T)$. We know that $\Delta(T', OS(X)) = \Delta(T, OS(X)) - 1$. If $y \in SL(T')$ then the same argument would lead to $\Delta(T, \mathcal{OS}(X)) = \Delta(T', \mathcal{OS}(X)) - 1$ for the simple reason that $T' = T'_{<x, y>} = T$. Therefore $y \notin SL(T')$. ■

The following property does not hold for the Copeland and Markov methods.

Proposition 3.4.8.: The Slater solution is weakly composition-consistent.

Proof :

Let Y be a component of $T \in \mathcal{T}(X)$, $x, y \in Y$, $x \neq y$. Denote $T' = T_{<x, y>}$. Let $z \in SL(T)$, $z \notin Y$. There exists U, a Slater order for T with z on top. The restriction of U to $X - \{z\}$ is a Slater order for the restriction $T\text{-}z$ of T to $X - \{z\}$. The tournament $T - z$ has also Y as a component, so that, considering a Slater order for $T - z$ such that Y is an interval, we may suppose that Y is an interval for U. Hence Y is a component of both T and U, and one has :

$$\begin{aligned}\Delta(T, U) = \Delta(T/(X\text{-}Y), U/(X\text{-}Y)) + \Delta(T/Y, U/Y) \\ +(\#Y).\ \#\{x \in X : xTy \text{ and } YUx\} \\ +(\#Y).\ \#\{x \in X : yTx \text{ and } xUY\}.\end{aligned}$$

Changing T to T', one can easily find a Slater order U' for T' by reordering in U the interval Y. Thus $z \in SL(T')$; this is sufficient to prove :

$$SL(T') \cap (X - Y) = SL(T) \cap (X - Y).$$

If $Y \cap SL(T) \neq \emptyset$ then the same reasoning shows that there exists a Slater order for T' obtained by first choosing a Slater order for T with Y on top and then reordering the interval Y according to a Slater order for T'/Y. Therefore :

$$SL(T) \cap Y \neq \emptyset \Rightarrow SL(T') \cap Y \neq \emptyset.$$

and the proof is complete. ■

4 - Multivariate Description

This chapter is devoted to the use of multivariate description methods in the case of tournaments. We present two methods of geometric description. The aim of these methods is to find "optimal" (in a sense to be made precise) embedding of the vertices of a tournament into some low-dimension euclidean space.

Given a tournament T on X and three alternatives x, y and z in X, we say that x and y are identical with respect to z if x and y both beat z or if they are both beaten by z. On the other hand, x and y are very different if for many alternatives z we have $xTzTy$ or $yTzTx$. So consider the quantity:

$$e(x,y) = \#\{z \in X : z \neq x,\ z \neq y \text{ and } xTzTy \text{ or } yTzTx\}.$$

With e, we have not defined yet a distance on X since if $\{x,y\}$ is a component of T then $e(x,y)=0$ even if $x \neq y$, but it is easy to modify e in order to get an actual distance. In fact, there are several ways to do so, which lead, as we shall see, to different methods. Suppose that we have defined such a distance, d, on X. Then we can use data analysis techniques in order to find an image of X in a low-dimension vector space for which the euclidean distances between the images of the points of X are as close as possible to the distances $d(x,y)$.

4.1. Complete Euclidean Description

In fact there is a natural euclidean space which the vertices of a tournament lie in. Let T be tournament on a set X of n alternatives, and consider the n-dimensional vector space $\mathbb{R}^X$, in this space each axis refers to an alternative of X. The tournament T is represented in $\mathbb{R}^X$ by a set of n points, one for each alternative, in the following way : the coordinate of a point $x \in X$ on the axis $i \in X$ is +1 if xTi, -1 if iTx and 0 if $x=i$. This coordinate will be denoted by x_i. So the reader will have to keep in mind that, in this representation, the alternatives of the tournament are simultaneously the original axes of the space and points in the same space. In order to avoid confusion, we shall use the letters x, y, ... to denote the points and the letters i, j, ... to denote the axes, and we call "variables" the directions of the space. The square matrix N whose entry is x_i on the row x and the column i records the coordinates of the points $x \in X$ in the *original basis* $\{i \in X\}$. It is an antisymmetric ("skew-symmetric") matrix : $N^T = -N$. The matrix N will be called the *comparison matrix* of the tournament T. If M is the tournament matrix then $N = M - M^T$. For instance, if T is a transitive tournament of order 3, the comparison matrix is :

$$\begin{matrix} 0 & +1 & +1 \\ -1 & 0 & +1 \\ -1 & -1 & 0 \end{matrix}$$

The tournament is a set of three points in a three dimensional space, which forms a triangle ; in the original basis this triangle is *not* the simplex. Observe also that the three points are not on a line, despite the fact that the tournament is here a so-called *linear* ordering.

Now we endow the vector space $\mathbb{R}^X$ with the usual euclidean structure and denote by E_2 this normed space. The distance between two points x and y of E_2 is:

$$d_2(x,y) = (\sum_{i \in X} (x_i - y_i)^2)^{1/2}.$$

The scalar product will be denoted by <x,y>. Using matrix notations, we write the points as row vectors, so that we have:

$$<x,y> = x.\, y^T = \sum_{i \in X} x_i y_i.$$

The norm of x is denoted by $\|x\|$ and $\|x\|^2 = <x,x>$. If x and y are in X, we have nice interpretations of these numbers: the product $x_i y_i$ equals 0 if $x=i$ or $y=i$, it equals +1 if x and y are in the same position with respect to i (that is to say if xTi and yTi or if iTx and iTy) and it equals -1 if x and y are not in the same position with respect to i (that is to say if xTi and iTy, or if iTx and yTi). Similarly, $(x_i - y_i)^2$ takes the values 0 if x and y are in the same position with respect to i and $2^2=4$ if they are not; it takes the value 1 if $x=i$ and $y \neq i$ or if $x \neq i$ and $y=i$. So two alternatives are close to each other if they beat the same other alternatives. The distance d_2 is related to the quantity e previously introduced by the formula:

$$d_2(x,y)^2 = 2 + 4e(x,y).$$

Observe also that, in the tournament case, all the points of X have the same norm: $\|x\|^2 = n-1$, so that they all are located on a sphere centered at the origin.

In order to describe the tournament T, we consider the system of points X as a solid body (or "cloud of points") in the n-dimensional euclidean space and we look for optimal representations of this body in lower dimension. The technique for doing so is the Principal Component Analysis. In the sequel, we shall recall this technique without proofs, except for the properties which are specific to the analysis of antisymmetric matrices. For more information about principal components, we refer the reader to any textbook in statistics, for instance Volle (1985) or Anderson (1984). In the vocabulary of statistics, we make a centered and not normed analysis, that is to say that we analyze the body T from its center of gravity and without changing the metric structure (which here has a good intuitive meaning).

The *center of gravity* of T is the point β whose coordinate β_i is:

$$\beta_i = (1/n) \sum_{x \in X} x_i.$$

It is easy to see that in the antisymmetric case:

$$\sum_{i \in X} \beta_i = 0.$$

In matrix notations, let **1** denotes the column vector of 1, then:

$$\beta = (1/n)\mathbf{1}^T.N.$$

Let J denotes the square matrix with $1/n$ in each of its entries, the matrix, denoted by N', of the centered cloud has $x\text{-}\beta$ on each row x, that is: $N' = N - J.N$. The *matrix of inertia* of T, denoted by G, is then by definition the symmetric matrix:

$$G = N'^T.N'.$$

The total *inertia* of the body T is the positive quantity:

$$In = (1/2) \sum_{x,y \in X} \|x - y\|^2 = \sum_{x \in X} \|x - \beta\|^2$$

and the inertia of T along some affine subspace F of E_2 is simply the inertia of the projection of T on F, it is called the inertia *explained* by F. A *best representation* of dimension $m \leq n$ of T is a projection on an affine subspace of dimension m with largest explained inertia. Intuitively, it is a way to represent the n-dimensional body in an m-dimensional space such that the distances are preserved, as much as possible. It is geometrically clear, and easy to prove, that the center of gravity belongs to all the best representations. Of special interest is the case where F is a line passing by β: $F(\alpha) = \{k\alpha + \beta, k \in \mathbb{R}\}$ for some $\alpha \in E_2$ that we can suppose normed: $\|\alpha\| = 1$. Then the projection of $x \in X$ on $F(\alpha)$ is the vector $\langle x\text{-}\beta, \alpha \rangle \alpha + \beta$ and the quantity $\langle x\text{-}\beta, \alpha \rangle$ is generally referred to as the *score* of x along α; we shall keep this denomination. The inertia explained by $F(\alpha)$ is called the inertia in the direction α and can be written as:

$$In(\alpha) = \sum_{x \in X} <x-\beta,\alpha>^2 = \alpha^T.G.\alpha \qquad (1)$$

(α is a column vector).

The matrix G being symmetric, it has positive eigenvalues $\lambda_1 \geq \lambda_2 \geq ... \geq \lambda_n \geq 0$. The fundamental theorem of the Principal Components Analysis is then:

<u>Theorem 4.1.1</u>: A best m-dimensional representation of T is the projection of T on an affine subspace passing by the center of gravity and directed by m eigenvectors of the matrix of inertia G associated to the m largest eigenvalues of G.

The best m-dimensional representation is not necessarily unique since the eigenvalues are not necessarily distinct. We can find an orthonormal basis $\{\alpha^1, \alpha^2,... \alpha^n\}$ of eigenvectors for G, α^u being associated with λ_u, for u=1, ... n. These directions are called the *principal components* of T. The inertia explained by component α^u is precisely the associated eigenvalue:

$$In(\alpha) = \lambda_u.$$

The base $(\beta; \alpha^1, \alpha^2,... \alpha^n)$ of the affine space E_2 will be called the *principal base*. Let A be the matrix whose u-th column is α^u, the inverse of A is its transpose and $\Lambda = A^T.G.A$ is the diagonal matrix of the eigenvalues. The score of x along α^u is :

$$k^u(x) = <x-\beta,\alpha^u> = (x-\beta).\alpha^u, \qquad (2)$$

clearly:

$$\sum_{x \in X} k^u(x) = 0.$$

The scores of the various points of X along the principal components are given by a matrix K which we shall call the *score matrix*:

$$K = N'.A$$

(each row of K gives the scores of a point of X). Note that even if all the eigenvalues are distinct, the scores are defined up to a multiplication by -1: in general there is no reason to prefer an orientation to the other. We shall see that in the case of a tournament there is a natural way to orient half of the principal components. From $\Lambda = A^T.G.A$, $G=N'^T.N$ and $K = N'.A$ one deduces:

$$A.\Lambda = N'^T.K,$$

that is to say, for all u:

$$\lambda_u \alpha^u = N'^T.k^u. \qquad (3)$$

We can give interpretations of the equations (2) and (3). Consider an axis $u \in \{1, ..., n\}$. The equations can be written:

$$\text{for all } x \in X : k^u(x) = \sum_{i \in X} \alpha^u{}_i,$$

$$\text{for all } i \in X : \lambda_u \ \alpha^u{}_i = \sum_{x \in X} k^u(x).$$

Using the fact that $\sum_{x \in X} k^u(x) = 0$, one gets:

$$\text{for all } x \in X\text{: } k^u(x) = \sum_{i:xTi} \alpha^u{}_i - \sum_{i:iTx} \alpha^u{}_i - \sum_{i \in X} \beta_i \alpha^u{}_i \qquad \text{(2bis)}$$

$$\text{for all } i \in X\text{: } \lambda_u \ \alpha^u{}_i = \sum_{x:xTi} k^u(x) - \sum_{x:iTx} k^u(x). \qquad \text{(3bis)}$$

Call $\alpha^u{}_i$ the *importance* of i, then (3bis) states that the importance of $i \in X$ is (up to the multiplicative constant λ_u) the total score of the alternatives that i beats minus the total score of the alternatives that are beaten by i. The other equation (2bis) states that (up to the additive constant $-\sum_{i \in X} \beta_i \alpha^u{}_i$) the score of $x \in X$ is the total importance of the alternatives that x beats minus the total importance of the alternatives that are beaten by x.

Let us already take an example. We consider the tournament T of order 7 depicted in figure 4.1. The vertices are: $X=\{1,2,3,4,5,6,7\}$ and the reader will verify that the order shown in figure 4.1 is a Slater order for T. This example has been chosen because it is not too far from a linear order (three arrows to be reversed) but different scoring methods gives different results, as will be shown. In a sense, this tournament is neither pathological nor too simple (results for this tournament are summarized in the annex).

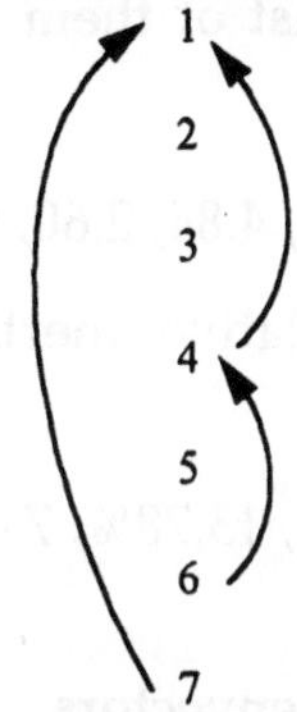

Figure 4.1.

The tournament matrix is:

$$M = \begin{matrix} 0 & 1 & 1 & 0 & 1 & 1 & 0 \\ 0 & 0 & 1 & 1 & 1 & 1 & 1 \\ 0 & 0 & 0 & 1 & 1 & 1 & 1 \\ 1 & 0 & 0 & 0 & 1 & 0 & 1 \\ 0 & 0 & 0 & 0 & 0 & 1 & 1 \\ 0 & 0 & 0 & 1 & 0 & 0 & 1 \\ 1 & 0 & 0 & 0 & 0 & 0 & 0 \end{matrix}$$

The comparison matrix $N = M - M^T$ is:

$$\begin{matrix} 0 & +1 & +1 & -1 & +1 & +1 & -1 \\ -1 & 0 & +1 & +1 & +1 & +1 & +1 \\ -1 & -1 & 0 & +1 & +1 & +1 & +1 \\ +1 & -1 & -1 & 0 & +1 & -1 & +1 \\ -1 & -1 & -1 & -1 & 0 & +1 & +1 \\ -1 & -1 & 1 & +1 & -1 & 0 & +1 \\ +1 & -1 & -1 & -1 & -1 & -1 & 0 \end{matrix}$$

The center of gravity has coordinates:

$$\beta = (-2/7, -4/7, -2/7, 0, 2/7, 2/7, 4/7).$$

Diagonalisation of the matrix of inertia leads to 7 different eigenvalues, λ_1, ..., λ_7, the last of them being zero, $\lambda_7=0$. Those are approximately:

$$\lambda = (15.86, 11.30, 4.84, 2.60, 0.30, 0.24, 0).$$

It is worth quoting the relative inertia explained by the seven principal components:

$$(45.12\%, 32.16\%, 13.78\%, 7.40\%, 0.85\%, 0.68\%, 0\%).$$

The first three eigenvectors associated with the first eigenvalues of the matrix of inertia G can be taken as:

$$\alpha^1 = (-0.371, 0.329, 0.524, 0.209, 0.404, 0.519, -0.056)^T$$
$$\alpha^2 = (0.437, 0.380, 0.227, -0.564, 0.126, -0.084, -0.522)^T$$
$$\alpha^3 = (0.470, -0.030, 0.144, 0.431, 0.605, -0.415, 0.183)^T$$

(They are normalized so that they have unit norm).

The coordinates of the seven points on the three first principal axes are given by the first three columns of the score matrix K:

point 1:	(1.623, 2.427, -0.310)
point 2:	(1.971, -0.561, 0.477)
point 3:	(1.118, -1.168, 0.363)
point 4:	(-1.395, 0.213, 1.558)
point 5:	(-0.228, -0.394, -1.248)
point 6:	(-0.733, -1.563, -0.575)
point 7:	(-2.356, 1.046, -0.265)

Pictures 4.2. and 4.3. show the projections of the tournament on the planes (α^1,α^2) and (α^1,α^3) respectively, with the origin at the center of gravity.

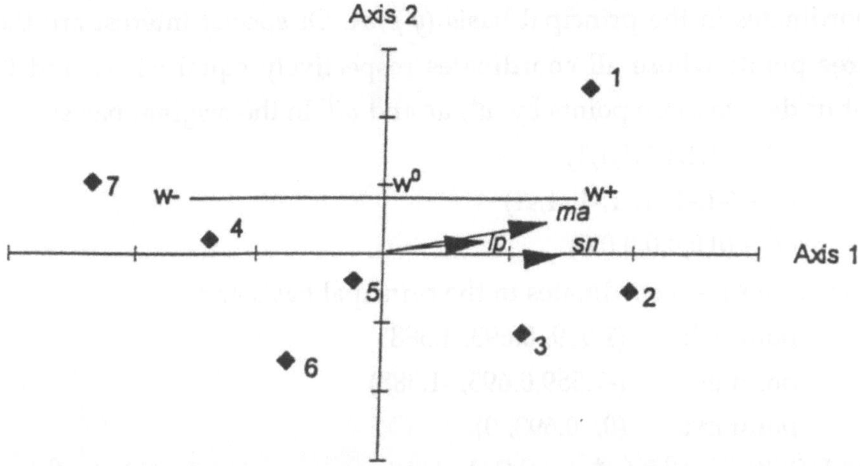

Figure 4.2.

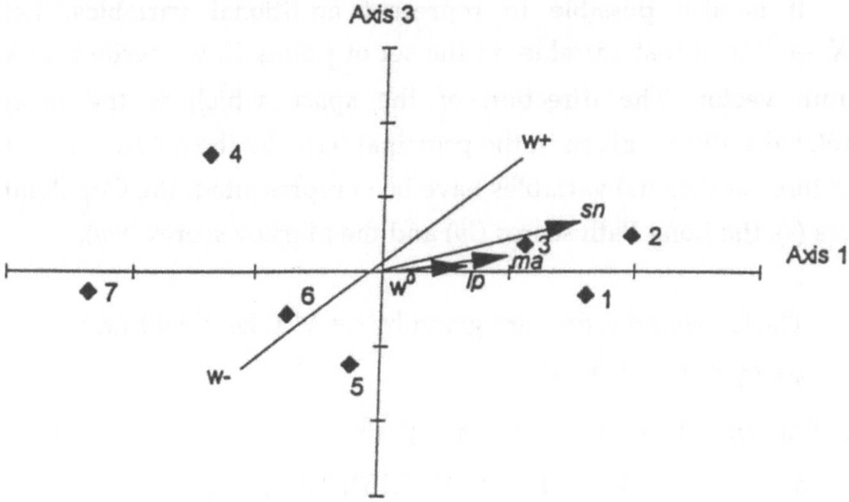

Figure 4.3.

Additional points and variables.

A point $y \in \mathbb{R}^X$ of coordinates y_i $(i \in X)$ in the original basis has coordinates in the principal basis $(y-\beta).A$. Of special interest are the three points whose all coordinates respectively equal +1, -1 and 0. Let us denote these points by ω^+, ω^- and ω^0. In the original basis:

$\omega^+ = (1,1,1,1,1,1,1)$

$\omega^- = (-1,-1,-1,-1,-1,-1,-1)$

$\omega^0 = (0,0,0,0,0,0,0)$.

Their first three coordinates in the principal basis are:

point ω^+: (1.559, 0.693, 1.388)

point ω^-: (-1.559,0.693, -1.388)

point ω^0: (0, 0.693, 0).

By definition, $\omega^0 = (\omega^+ + \omega^-)/2$, but it is a surprising fact that ω^0 has coordinate 0 on several axes and that on the other axes the three points have the same coordinates. This fact is not particular to the example, and it will be explained later.

It is also possible to represent additional variables. Let $v : X \to \mathbb{R}$ be a real variable on the set of points X, we write v as a column vector. The direction of the space which is the most correlated with v is given in the principal basis by the column vector $K^T.v$; three additional variables have been represented, the Copeland scores (s), the Long Path scores (lp) and the Markov scores (ma).

The Copeland scores are given by $s = M.1$, here we find :

$s = (4, 5, 4, 3, 2, 2, 1)^T$.

and after normalization, letting $sn = s/\|s\|$:

$sn = (0.462, 0.577, 0.462, 0.346, 0.231, 0.231, 0.115)$.

For the Long Path scores, we compute a Perron eigenvector of the tournament matrix M. After normalization we find :

$lp = (0.533, 0.547, 0.391, 0.366, 0.176, 0.230, 0.212)^T$.

In order to compute the Markov scores, we compute an eigenvector of the transition matrix associated to T (see 3.2) and we find the (normed) vector:

$ma = (0.561, 0.738, 0.246, 0.241, 0.050, 0.0088, 0.112)^T$.

Observe that the rankings induced by these three scoring procedures are different, the three rankings are:

Copeland: 2, (1 and 3), 4, (5 and 6), 7.
Long Path: 2, 1, 3, 4, 6, 7, 5.
Markov: 2, 1, 3, 4, 7, 5, 6.

They also differ from the three Slater orders:

Slater: 1, 2, 3, 4, 5, 6, 7.
1, 2, 3, 5, 6, 4, 7.
1, 2, 3, 6, 4, 5, 7.

On the first three principal components, the additional variables are represented by the first three components of the column vectors $K^T.sn$, $K^T.lp$ and $K^T.ma$, these coordinates has been used to draw the figures 4.2. and 4.3. Here we give the complete representations of these variables:

Copeland: (1.427, 0, 0.388, 0, 0.012, 0, 0).
Long Path: (0.473, 0.164, 0.163, 0.046, 0.006, 0.001, 0).
Markov: (0.964, 0.329, 0.245, 0.125, 0.059, -0.006, 0).

On the two first axes, the directions of the variables Long Path and Markov are almost identical. One can observe that the correlation between the additional variables and the last principal component is zero; this is because the last eigenvalue (λ_7) is zero. More surprisingly, the correlation of the Copeland score with the second, fourth and sixth principal components also equals zero. In order to explain this fact, we need to go a little further in the mathematical analysis of antisymmetric matrices.

Principal Components of antisymmetric data.

Let N be any antisymmetric real matrix and consider the matrix of inertia, $G = (N - J.N)^T.(N - J.N)$, since $N^T = -N$, $J^T=J$ and $J^2=J$, one finds that:

$$G = -N^2 + N.J.N.$$

Observe that the matrix $J.N$ has all its rows equal to β. Let α be a (column) vector, then $J.N.\alpha = <\beta,\alpha>\mathbf{1}$, and since $N.\mathbf{1} = -n\beta$, one finds that for any α:

$$G.\alpha = -N^2.\alpha - n<\beta,\alpha>\beta. \qquad (4)$$

With this formula, we shall be able to relate the analysis of G to the one of the antisymmetric matrix N. Here are the useful facts about the linear applications defined by antisymmetric (skew-symmetric) matrices. For more information, we refer the reader to the chapters "Symplectic geometry" or "Skew mappings" of textbooks in algebra, for instance Artin (1978) or Greub (1981).

Proposition 4.1.2.: The eigenvalues of N are pure imaginary.

Proof :

Since N is normal ($N^*.N = N.N^*$, where N^* is the conjugate transpose of N), N is diagonalisable in $\mathbb{C}$ by a unitary matrix V. (V is such that $V^*.V=I$.) Let v be a (complex) eigenvector of N associated to the (complex) eigenvalue ν : $Nv=\nu v$. then $-N^2=-\nu^2 v$, but $-N^2=N^T.N$ shows that $-N^2$ is (real) symmetric, and so its eigenvalues are real ≥ 0. Hence ν is pure imaginary. ■

So we can write the eigenvalues of N: $\nu=i\mu$, with $\mu \in \mathbb{R}$. Eventually changing v to its conjugate, it is possible to take $\mu \geq 0$. Let us write $v = v' + iv''$, v' and v'' the real and imaginary parts of vector v. Let $v^*=v'-iv''$ denote the conjugate of v; clearly: $Nv=i\mu v$ if and only if $Nv^*=-i\mu v^*$. The proper subspaces associated to the eigenvalues $i\mu$ and $-i\mu$ are conjugate and, in particular, they are of the same dimension. Eventually changing v to its conjugate, it is possible to take $\mu \geq 0$. We choose the unitary matrix V so that:

$$V = (v^1, v^{1*}, w^1, w^{1*}, \ldots, v^2, v^{2*}, \ldots),$$

with $v^1, w^1, \ldots$ associated to the eigenvalue $i\mu_1$, $v^{1*}, w^{1*}, \ldots$ associated to $-i\mu 1$, $v^2, \ldots$ associated to $i\mu_2$, and so on. Let $[v',v'']$ denote the real vector space spanned by the real vectors v' and v''. One has:

<u>Proposition 4.1.3.</u>: Let v be a (complex) unitary eigenvector of N associated with the eigenvalue $i\mu$, if $\mu \neq 0$ then $[v',v'']$ is a plane, v' and v'' are orthogonal, $\|v'\|^2 = \|v''\|^2 = 1/2$ and the restriction of the real linear application N to $[v',v'']$ is a similitude of angle $-\pi/2$ in the basis (v',v'') and of scale μ.

Proof :

Taking the real and imaginary parts of the equation $N.v = i\mu v$ one finds:

$$N.v' = -\mu v''$$
$$N.v'' = \mu v', \tag{5}$$

so that:

$$-N^2.v' = \mu^2 v'$$
$$-N^2.v'' = \mu^2 v'',$$

and also:

$$v''^T.N.v' = -\mu \|v''\|^2$$
$$v'^T.N.v'' = \mu \|v'\|^2.$$

Because N is antisymmetric, $v''^T.N.v' = -v'^T.N.v''$, so if $\mu \neq 0$,

$$\|v'\|^2 = \|v''\|^2.$$

Since v is unitary, $v.v^* = 1$, that is to say:

$$\|v'\|^2 + \|v''\|^2 = 1, \text{ and so,}$$
$$\|v'\|^2 = \|v''\|^2 = 1/2.$$

It is easily checked that $v' \neq v''$ and that $v' \neq -v''$, so $[v',v'']$ is a plane. Since N is antisymmetric, $0 = v'^T.N.v' = -\mu <v',v''>$, so v' and v'' are orthogonal. Thus (3) shows that the restriction of N to the real plane

$[v',v'']$ is a similitude of angle $-\pi/2$ and of scale μ, and the proof is complete. ■

Let us call *proper plane* of N a plane $[v',v'']$ corresponding to a complex eigenvector of N (or its conjugate) associated to a non-zero eigenvalue. The last needed result is easily deduced from the fact that V is unitary.

Proposition 4.1.4.: Two proper planes are orthogonal.

We are now in position to understand a little better what happens to the principal component analysis of antisymmetric data.

Let α be a unitary vector of $\mathbb{R}^X$, according to (1) and (4), the inertia in the direction α can be written:

$$In(\alpha) = -\ \alpha^T.N^2.\alpha - n <\beta,\alpha>^2.$$

In particular, $In(\alpha)$ is lower than $-\ \alpha^T.N^2.\alpha$, and equals this quantity if α is orthogonal to β. But of course, within each proper plane of N it is possible to find an α orthogonal to β, so we have:

Proposition 4.1.5.: Let $\mu^2 \neq 0$ be an eigenvalue of $-N^2$, then μ^2 is an eigenvalue of G and there exists an eigenvector α of G associated with μ^2 such that $<\beta,\alpha>=0$.

This proposition will be useful after two following ones:

Proposition 4.1.6.: The scores of the points ω^+, ω^- and ω^0 along a direction α are respectively:

$$\mathbf{1}^T.\alpha - <\beta,\alpha>,$$
$$-\mathbf{1}^T.\alpha - <\beta,\alpha>,$$
$$- <\beta,\alpha>.$$

Proof:

These scores are simply $<\omega^+ - \beta, \alpha>$, $<\omega^- - \beta, \alpha>$ and $<\omega^0 - \beta, \alpha>$. ■

So along those directions such that $<\beta,\alpha>=0$, we find that α^0 has zero score, as was seen on the example.

Proposition 4.1.7.: The correlation between the Copeland score and a principal component α associated to the eigenvalue λ is: $(\lambda/2)\, 1^T.\alpha$.

Proof :

Recall that the score matrix K is: $K=N'.A$ and that on the principal component α, $\lambda\alpha = N'^T.k$. So for all $i \in X$:

$$\sum_{x\in X} (x_i-\beta_i)k(x) = \lambda\, \alpha_i.$$

Summing on i and using the fact that $\sum_{x\in X} k(x) = 0$, one finds:

$$\sum_{x\in X}\sum_{i\in X} x_i k(x) = \lambda \sum_{i\in X} \alpha_i.$$

But $\sum_{i\in X} x_i = -n\, \beta_x = 2s(x) -n + 1$, so we find $\sum_{i\in X} 2s(x)k(x) = \lambda \sum_{i\in X} \alpha_i$ and the result. ■

In the example, we have chosen the orientations of the axes so that $1^T.\alpha\geq 0$, in order to have positive correlations with the Copeland score. Call *Copeland principal components* the principal components correlated with the Copeland score. Those components appear in the decomposition of the space according to the proper planes of N. For any proper plane $[v',v'']$, consider the projection ε of β on this plane. Suppose that $\varepsilon\neq 0$ and let $a'= \varepsilon/\|\varepsilon\|$. Then, according to the proposition 4.1.3, $Na'= -\mu\, a''$ for a vector $a''\in[v',v'']$ such that (a',a'') is an orthonormal basis of $[v',v'']=[a',a'']$. One also has $Na''= \mu a'$. By definition of a' and a'', $<\beta,a''>=0$, so $Ga''=\mu^2 a''$. The action of N on the vector 1 is already known: $N1=-\beta$, so within the proper plane $[a',a'']$, $<1,a'>=0$. From this will be deduced :

Proposition 4.1.8.: For any Copeland principal component α^u, $<\beta,\alpha^u>=0$.

Proof :

Let α^u be a Copeland principal component, consider the orthonormal basis $(\alpha^1, \ldots, \alpha^n)$ of eigenvectors of G and the associated eigenvalues $(\lambda_1, \ldots, \lambda_n)$. Among these eigenvalues we find all the eigenvalues μ^2 of $-N^2$ and among these eigenvectors we find all the eigenvectors (denoted by a'') of $-N^2$ orthogonal to β. Any other α belongs to the vector space spanned by vectors a' such that $<\mathbf{1},a'>=0$ and by eigenvectors of G associated to $\lambda=0$, so if $\lambda_u \neq 0$ and $<\mathbf{1}, \alpha^u> \neq 0$ then $<\beta,\alpha^u>=0$ and the result follows from the proposition 4.1.7. ■

In view of the proposition 3.1.6, the previous proposition means that the point $\omega^0 = (0,0, \ldots ,0)$ has zero score on any Copeland principal component. Or more geometrically: along any Copeland principal component, the origin is projected on the center of gravity.

Proposition 4.1.9.: There are at most $En[n/2]$ Copeland principal components.

Proof :

Consider a proper plane $[v',v'']$ and the corresponding eigenvalue $i\mu$ of N. If the projection of β on $[v',v'']$ is $\neq 0$ then we have already noticed that there is a direction $a' \in [v',v'']$ such that $<\mathbf{1},a'>=0$, on the other hand, if β is orthogonal to $[v',v'']$, then we have:

$$\beta.v' = \beta.v'' = 0.$$

But $\beta = \mathbf{1}^T.N$, so equation (5) leads to:

$$\mu\mathbf{1}^T.v' = \mu\mathbf{1}^T.v'' = 0$$

and thus $\mu\mathbf{1}^T$ is orthogonal to the whole plane $[v',v'']$. This shows that each proper plane gives at least one, and maybe two, directions which have zero correlation with the Copeland score. The proposition is then easily deduced from the proposition 4.1.4. ■

Some other curious facts about tournaments can be seen in that geometric representation. For instance, since a tournament T of order n is made of n points in an euclidian space, it is included in an affine space of dimension $\leq n$-1, thus there exists a vector γ such that $N'\gamma=0$. Clearly $G\gamma=0$ thus the last eigenvalue of G is always 0. the equation $N'\gamma=0$ can be written : $N\gamma = \langle\beta,\gamma\rangle 1$, or more clearly : for any vertex $x \in X$

$$\sum_{y:xTy} \gamma_y - \sum_{y:yTx} \gamma_y = constant.$$

In the example we find for the last principal component:

$$\alpha^7 = (0.6255, 0.2085, -0.2085, 0.2085, -0.2085, 0.6255, 0.285)^T,$$

and we can take:

$$\gamma = (3, 1, -1, 1, -1, 3, 1)^T,$$

the constant $\langle\beta,\gamma\rangle$ being 0. Thus, given any tournament, one can associate (positive or negative) weights γ_x to its vertives so that for every vertex x the difference between the total weight of its predecessors and the total weight of its successors is constant. Here we find a kind of equilibrating system, somehow in the spirit of those studied by Moon and Pullman (1970), but no interpretation of these weights γ_x is known.

4.2. Multidimensional Scaling

The method of representation which has been explained in the first part of this chapter has the property that the linear orderings (of order >2) are not represented by points on a line. This is due to the use of the distance

$d_2(x,y) = (2 + 4e(x,y))^{1/2} = (\sum_{i \in X}(x_i - y_i)^2)^{1/2}.$

If one considers the distance:

$d_1(x,y) = 1 + 2e(x,y)$

which is simply the distance of the symmetric difference, it is easily seen that linear orderings are perfectly represented by putting the points regularly on a line, the number $e(x,y)$ being the number of outcomes between x and y. So it would be interesting to have a good euclidean representation of the array of distances $(d_1(x,y))_{x,y \in X}$.

But, with the distance d_1, it is not the case that any tournament can be embedded in an euclidean space, even of high dimension (linear orders are notable exceptions). Nevertheless there exists a technique for finding euclidean approximate representations of non-euclidean sets of distant points. This technique, known as Multidimensional Scaling or "analysis of distance tableaux" is based on the observation that, in a principal components analysis, the eigenvalues and the scores of the points on the principal components can be computed knowing only the distances between the points (and not their initial coordinates). This fact will not be demonstrated here (see for instance Volle 1985 or Seber 1984) and we give the formulae without proofs.

Given is a set X of n points and a distance d on X. Let D be the symmetric (n,n) matrix giving the squares of the distances in X:

$D_{x,y} = (d(x,y))^2$

and let Γ be the matrix:

$$\Gamma = (1/2)\ (D.J + J.D - D - J.D.J). \qquad (6)$$

If $d = d_2$, then Γ is not the matrix of inertia $G = N'^T.N'$ previously studied but the matrix of inertia of the dual analysis, $\Gamma = N'.N'^T$; but Γ and G have the same eigenvalues: let λ be a non-zero eigenvalue of Γ and w be a λ-eigenvector of Γ, then $\Gamma w = \lambda w$ implies $G(N'^T w) = \lambda(N'^T w)$, thus $N'^T w$ is an eigenvector of G associated to the eigenvalue λ. One has:

$$||N'^T w||^2 = w^T(N'N'^T)w = \lambda||w||^2.$$

Hence if w is a normed λ–eigenvector of Γ then $\alpha = (\lambda^{-1/2})\, N'^T w$ is a normed λ–eigenvector of G. The vector of the scores along α is $k^\alpha = N'\alpha$, which can be written:

$$k^\alpha = (\lambda^{-1/2})\, \Gamma w = (\lambda^{1/2})\, w.$$

Therefore knowing Γ is sufficient for computing the scores k. In practice, one computes the matrix D with the distance d_1 :

$$D_{x,y} = (\, d_1(x,y)\,)^2 = (1 + 2e(x,y)\,)^2$$

and the matrix Γ with formula (6), and diagonalises Γ. Let $(\lambda_1, ..., \lambda_m)$ be the m first eigenvalues of Γ and let $(w^1, ..., w^m)$ be associated normed eigenvectors. In the m-dimensional euclidean space spanned by $(w^1, ..., w^m)$, a point $x \in X$ has coordinates $(k^1(x),.., k^m(x)\,)$, with $k^u = (\lambda_u{}^{1/2})\, w^u$ $(1 \leq u \leq m)$. Of course, this representation is only valid in the case where the first values are non-negative.

Let us take as an illustration the trivial case of a linear order on three points. Take the tournament matrix:

$$M = \begin{matrix} 0 & 1 & 1 \\ 0 & 0 & 1 \\ 0 & 0 & 0 \end{matrix}$$

the distance array is:

$$\begin{matrix} 0 & 1 & 2 \\ 1 & 0 & 1 \\ 2 & 1 & 0 \end{matrix}$$

thus the matrices D and Γ are:

$$D = \begin{matrix} 0 & 1 & 4 \\ 1 & 0 & 1 \\ 4 & 1 & 0 \end{matrix}$$

$$\Gamma = \begin{matrix} 1 & 0 & -1 \\ 0 & 0 & 0 \\ -1 & 0 & 1 \end{matrix}$$

The only non-zero eigenvalue of Γ is $\lambda_1=2$ and the corresponding normed eigenvector is $w^1 = (2^{-1/2}, 0, - 2^{-1/2})$, the scores of the three points on the axis w^1 are $2^{1/2} w^1 = (+1,0,-1)$.

As a non-trivial example, consider the tournament of order 7 which has been studied earlier. The matrix of the squares of the distances is :

$$D = \begin{matrix} 0 & 1 & 4 & 9 & 9 & 25 & 25 \\ 1 & 0 & 1 & 16 & 9 & 9 & 36 \\ 4 & 1 & 0 & 9 & 4 & 4 & 25 \\ 9 & 16 & 9 & 0 & 9 & 9 & 4 \\ 9 & 9 & 4 & 9 & 0 & 4 & 4 \\ 25 & 9 & 4 & 9 & 4 & 0 & 9 \\ 25 & 36 & 25 & 4 & 4 & 9 & 0 \end{matrix}$$

The matrix $\Gamma = (1/2)\ (D.J + J.D - D - J.D.J)$ has eigenvalues :

$$\lambda = (26.20, 10.91, -5.79, 3.86, -3.06, 0.03, 0).$$

Some of these eigenvalues are negative, so we restrict our attention to the two first ones. On these two first axes, the scores of the seven points are:

$$k^1 = (1.921, 2.591, 1.464, -1.100, -0.617, -0.906, -3.353)$$
$$k^2 = (1.916, -0.557, -0.702, 1.154, -0.266, -2.158, 0.614).$$

The corresponding two-dimensional representation is given in figure 4.4. It is not very different from the representation obtained using the distance d_2 (compare figure 4.4 to figure 4.2). Three additional variables have been represented (Copeland, Long-Path and Markov), for instance the correlation ('covariation') between the Copeland score and the two axes are respectively $k^1 .s = 16.797$ and $k^2.s = 1.296$, and the ratio 1.296/16.797 gives the slope of the represented line.

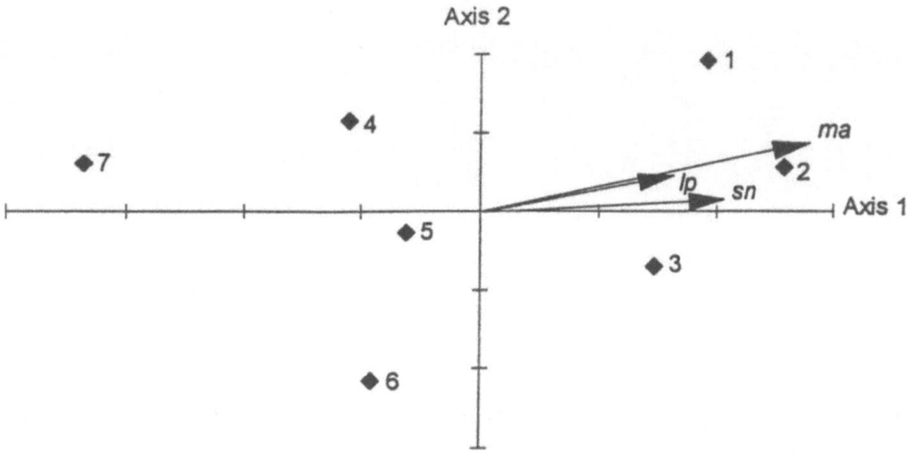

Figure 4.4.

5 - Covering

Methods of the previous chapters choose in tournaments by picking the maximal elements of some ranking (complete transitive binary relation) R associated to tournament T. Let R be a transitive binary relation associated to T; if T itself is not transitive the cost to pay for R being complete is that, for some x and y, xRy and yTx. In this chapter we define a transitive relation called the covering relation which is a sub-relation of T: if x covers y then xTy. Of course, such a relation cannot be complete unless T itself is transitive.

After the definition of the covering relation and of its associated tournament solution called the Uncovered set (5.1), we study refinements of the Uncovered set, its iterations (5.2), and the Minimal Covering set (5.3). Then we consider other solutions concepts obtained by slight modification of the covering relation (5.4 and 5.5).

An early reference to the uncovered alternatives is Landau (1953), and Fishburn (1977) considered the choice function defined by the Uncovered set. Following Maurer (1980), the elements of the Uncovered set are sometimes called the *kings* of the tournament, in the graph-theory literature (for instance Reid 1982 or Jing and Weixuan 1987). Properties of the Uncovered set are also studied by Bordes (1983), and the signifiance of the concept for Political Science is emphasized by McKelvey (1986). Extensions to the fuzzy framework are proposed by Perny (1994).

5.1. Covering Relation and Uncovered Set

Definition 5.1.1.: Let $T \in \mathcal{T}(X)$, $x, y \in X$. We say that x *covers* y in X if: xTy and $(\forall z \in X,\ yTz \Rightarrow xTz)$. When there can be no ambiguity about the tournament under consideration, we write : $x \Rightarrow y$.

The relation of covering is a sub-relation of T, thus asymmetric. It is transitive: let x, y and z be three points of X such that $x \Rightarrow y$ and $y \Rightarrow z$; then ($x \Rightarrow y$ and $y \rightarrow z$) implies that $x \rightarrow z$, and for any u in X such that $z \rightarrow u$, $y \Rightarrow z$ implies $y \rightarrow u$ and thus $x \Rightarrow y$ implies $x \rightarrow u$. Hence $x \Rightarrow z$, and the relation of covering is transitive, it is a *strict partial* ordering of X. Equivalent definitions of the covering relation can be given:

$x \Rightarrow y$ iff xTy and $\forall z \in X$, $T/\{x, y, z\}$ is transitive,

$x \Rightarrow y$ iff $x \neq y$ and $T^+(y) \subset T^+(x)$,

$x \Rightarrow y$ iff $x \neq y$ and $T^-(x) \subset T^-(y)$.

The reader will easily verify that these definitions are equivalent to 5.1.1. Following Miller (1980, 1983) we call "Uncovered set", and we denote by $UC(T)$, the set of maximal elements of this ordering. This set is non-empty since X is supposed to be finite. In the case where the covering relation is empty, that is to say when no point is covered, any point is a maximal element, and the Uncovered set equals X.

Definition 5.1.2.: The *Uncovered set* of T is :

$$UC(T) = \{x \in X : \nexists y \in X, y \Rightarrow x\}.$$

Observe that, in the tournament case, the definition of the covering relation also reads: x does not covers y if and only if either y beats x or there exists z such that y beats z and z beats x. Therefore another possible definition of the Uncovered set for tournaments is to say that $UC(T)$ is the set of points which beat any other point by a path of length one or two ("Two steps principle", Shepsle and Weingast 1984).

Proposition 5.1.3.: $x \in UC(T)$ if and only if :

$$\forall y \in X - \{x\}, xTy \text{ or } (\exists z \in X : xTz \text{ and } zTy).$$

Proof :

If $x \in UC(T)$, let $y \in X$ be different from x. If yTx then since y does not covers x, there exists $z \in X$ such that xTz and zTy. Conversely, if $x \notin UC(T)$ there exists y which covers x, that is yTx and for any $z \in X$, if xTz, yTz. ■

Remark 5.1.4.: The transitivity of the covering relation, and thus the possibility of the definition of an Uncovered set, hold for the oriented graphs which are not tournaments, but the preceding proposition essentially rests on the completeness of the tournament relation. Observe also that, if one takes for definition of the Uncovered set the proposition 5.1.3. it is no longer straightforward that this set is non-empty.

The Uncovered set is a solution concept, which can be defined by the following axiom, called expansion (or Sen's γ).

<u>Definition 5.1.5.</u>: A solution verifies the property of *expansion* if :
$\forall T \in \mathcal{T}(X), \forall Y \subset X, \forall Y' \subset X,$
$S(T/Y) \cap S(T/Y') \subset S(T/Y \cup Y').$

<u>Proposition 5.1.6.</u>: The Uncovered set is the finest tournament solution satisfying the expansion property.

Proof : (Moulin 1986)

Let us first prove that the Uncovered set satisfies the expansion property. Let Y and Y' be two subsets of X such that $UC(Y) \cap UC(Y') \neq \varnothing$ and take $x \in UC(Y) \cap UC(Y')$, then by the two-step principle, x can reach in two step any point of Y or of Y' thus $x \in UC(Y \cup Y')$.

Then let S be a tournament solution which satisfies the expansion property. If Y, Y' and Y'' are three subsets of X then

$$S(Y) \cap S(Y') \cap S(Y'') \subset S(Y \cup Y') \cap S(Y'') \subset S(Y \cup Y' \cup Y'')$$

and indeed, for any finite family $\{Y(z) : z \in Z\}$ of subsets of X :

$$\cap \{S(Y(z) : z \in Z\} \subset S(\cup \{Y(z) : z \in Z\}).$$

Consider $x \in UC(T)$. Denote $Y(x) = \{x\} \cup T^{+}(x)$. For each $y \in T^{-}(x)$, there exists u such that xTu and uTy ; denote $Y(y) = \{x, y, u\}$. Denote also $Z = \{x\} \cup T^{-}(x)$. Clearly, $X = \cup\{Y(z) : z \in Z\}$. Observe that $S(Y(x)) = \{x\}$ because x is a Condorcet winner for $T/Y(x)$ and that, for all $y \in T^{-}(x)$, $S(Y(y)) = Y(y)$ because $Y(y)$ forms a three cycle, thus $x \in \cap\{S(Y(z)): z \in Z\}$. Therefore $x \in S(T)$ and we proved that $UC \subset S$.■

Apart from the Top-Cycle, all the tournament solutions considered here are refinements of the Uncovered set. Hence, they do not satisfy the expansion property, The expansion property is satisfied by the Top-Cycle: Let $x \in TC\ (T/Y) \cap TC\ (T/Y')$, then from x one can reach any point in Y by some path in Y and any point in Y' by some path in Y', hence, $x \in TC\ (T/Y \cup Y')$. Here are some other properties of the Uncovered set.

<u>Theorem 5.1.7.</u>: The Uncovered set solution, *UC*

(i) is monotonous
(ii) is not independent of the losers
(iii) does not satisfies SSP
(iv) is not idempotent
(v) satisfies the Aïzerman property
(vi) is composition-consistent
(vii) is regular
(viii) is included in the Top-cycle.

Proof : (i), (vi), (vii) and (viii) are straightforward. For (ii), (iii) and (iv), consider the tournament *T* depicted in figure 5.1.

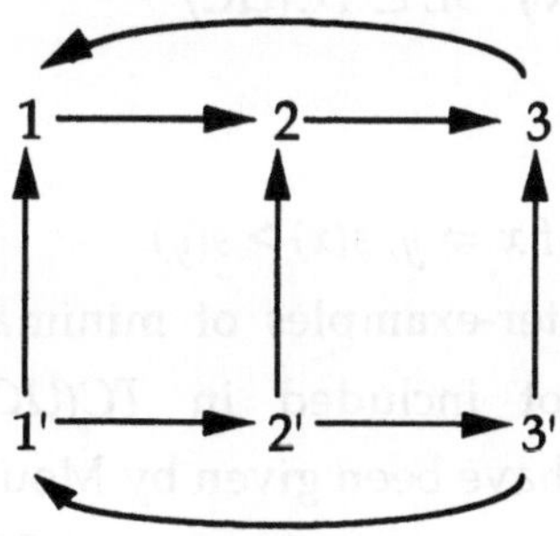

Figure 5.1.

It is not difficult to check that $UC(T) = \{1, 2, 3\}$. So consider the tournament $T' = T_{<1', 2'>}$ obtained from *T* by reversing the arrow $1' \to 2'$, one has $UC(T') = \{1, 2, 3, 2'\}$ thus *UC* is not independent of the losers. In the tournament restricted to $UC(T')$, the element 2' is a Condorcet loser, in particular, it does not belong to $TC(UC(T'))$, which proves that $TC(UC)$ is distinct from *UC* and that *UC* is not idempotent (point (iv)). The point (iii) is deduced from the point (iv).

Let us prove that *UC* satisfies the Aïzerman property: let $T \in \mathcal{T}(X)$, $Y \subset X$ with $UC(T) \subset Y$, we need to prove that $UC(T/Y) \subset UC(T)$. If $UC(T) = X$ it is obvious, if not, let $x \in X$, with $x \notin UC(T)$, there exists $y \in X$ which covers x in T; the covering relation being transitive, we may suppose that $y \in UC(T)$. Thus y belongs to Y, beats x and all the points beaten by x for T, which proves that $x \notin UC(T/Y)$, hence the result. ■

We now come to the relations that prevail between the Uncovered set and the scoring methods of the previous chapter. All these methods select subsets of the Uncovered set, but we can say slightly more:

Proposition 5.1.8.: (i) $C \subset UC$
(ii) $C \varnothing TC(UC)$
(iii) $Ma \subset UC$
(iv) $SL \subset TC(UC)$

Proof :

(i) In effect, if $x \Rightarrow y$, $s(x) > s(y)$

(ii) The counter-examples of minimal orders (10 and 13) such that $C(T)$ is not included in $TC(UC(T))$ and such that $C(T) \cap TC(UC(T)) = \varnothing$ have been given by Moulin (1986).

(iii) Let $x \Rightarrow y$, then $s(x) > s(y)$ and $\sum_{z \in T^+(x)} \overline{p}(z) \geq \sum_{z \in T^+(y)} \overline{p}(y)$, but the Markov scores are given by : $\overline{p}(x) = \frac{1}{o(T) - 1 - s(x)} \sum_{z \in T^+(x)} \overline{p}_z$, and likewise for y. So $\overline{p}(x) > \overline{p}(y)$ and the result.

(iv) This result was first proved by Banks, Bordes and Le Breton (1991). The following shorter proof is due to Le Breton (1996). We first prove that $SL \subset UC$. Let $x \in SL(T)$, write the alternatives x_1, x_2, ..., x_n according to a Slater order for T with $x=x_1$ on top. If x_1 was to be covered by some y then switching x_1 and y would stricly lower the distance to T, impossible. Thus $SL \subset UC$. We now prove that $SL \subset TC(UC)$. Let k be the first indice such that $x_k \in TC(UC(T))$. Suppose $k>1$. If $x_{k-1} \in UC(T)$ then x_{k-1} beats x_k and $x_k \in TC(UC(T))$ imply $x_{k-1} \in TC(UC(T))$, impossible. Thus x_{k-1} is covered in T by some point $x_{m'}$, which can be taken in $UC(T)$. The argument used to prove that $SL \subset UC$ shows that x_m covers x_{k-1} implies $m<k-1$. But x_m covers x_{k-1} in T and x_{k-1} beats x_k imply that x_m beats x_k. Thus $x_m \in TC(UC(T))$ contradicting the definition of k. Hence $k=1$ and the result. ■

The inclusions $C \subset UC$ and $Ma \subset UC$ give qualitative information about the Copeland and Markov winners. The next proposition shows that, in a certain way, one cannot do better.

Proposition 5.1.9. : The composition-expansions and composition-consistent hulls of the Copeland and Markov solutions are the Uncovered set :

$$\tilde{C} = C^* = \tilde{Ma} = Ma^* = UC .$$

Proof :

Because of propositions 2.5.4. and 5.1.8. and because UC is composition-consistent, $\tilde{C} \subset C^* \subset UC$ and $\tilde{M}a \subset Ma^* \subset UC$. Thus, we need to prove $UC \subset \tilde{C}$ and $UC \subset \tilde{M}a$. Let T be a tournament defined, without loss of generality on $\{1, ..., k\}$ and let $i \in U(T)$. For any odd integer n, consider the composed tournament $\tilde{T}^n$ defined by the expression :

$$\tilde{T}^n = \Pi(T; T_1, ..., T_k)$$

where, for each $j \in \{1, ..., k\}$, if jTi or $j = i$, T_j is the trivial tournament on $\{j\}$ and if iTj, T_j is a cyclone of order n.

We claim that, for n large enough, i is the unique Copeland and Markov winner of $\tilde{T}^n$. For $j \in \{1, ..., k\}$, denote:

$$a(j) = \#[T^+(j) \cap T^+(i)]$$

$$b(j) = \#[T^-(j) \cap T^+(i)]$$

Then $a(i) = \#T^+(i) = s(i)$ and $a(j) \leq a(i)$. If $a(j) = a(i)$ then $T(i) \subset T(j)$ and j covers i. Thus, $i \in UC(T)$ implies that $a(j) \leq a(i) -1$ for all $j \neq i$. Likewise, $b(j) > 0$ for all $j \neq i$.

For Copeland : The Copeland score of i in $\widetilde{T}^n$ is $n\,a(i)$. For j such that jTi, the Copeland score of j in $\widetilde{T}^n$ is

$$n\,a(j) + 1 + \#[\,T^+(j) \cap T^-(i)].$$

For any point in a component j such that iTj, the Copeland score in $\widetilde{T}^n$ is

$$n\,a(j) + \frac{n-1}{2} + \#[\,T^+(j) \cap T^-(i)].$$

Then it is easy to see that for n large enough, i is the unique Copeland winner of $\widetilde{T}^n$.

For Markov : The Markov scores in $\widetilde{T}^n$ can be computed with the formula given in proposition 3.3.6. Denote by q_j the sum of Markov scores in $\widetilde{T}^n$ of the points in component i. Then, if $j = i$ or if jTi, $\bar{p}_j = q_j$ and if iTj then, by symmetry, the Markov score of each point in the component i is $\frac{1}{n}q_j$. Thus :

For $j = i$: the Markov score of i is given by:

$$q_i = \frac{1}{s^-(i)} q_i^+$$

with $q_i^+ = \sum_{j \in T^+(i)} q_j^+$.

For j such that jTi:

$$(nb(j) + \#[T^-(j) \cap T^+(i)])\, q_j = \sum_{k \in T^+(i)} q_k$$

and because $b(j) > 0$: $q_j < \dfrac{1}{nb(j)}$.

Let $q_i^- = \sum_{j \in T^-(i)} q_j$, this last remark implies that there exists $A > 0$ such that $q_i^- < \dfrac{A}{n}$ for all n. But $q_i^- + q_i + q_i^+ = 1$ hence $q_i > \dfrac{1 - A/n}{1 + s^-(i)}$ and there exists $B > 0$ such that $q_i > B$ for all n. Therefore, the Markov score of any point in $\widetilde{T}^n$ except point i tends to 0 with $1/n$ when n tends to infinity and i is the unique Markov winner for n large enough.

We just proved that if $i \in UC(T)$ then $i \in \tilde{C}(T)$ and $i \in \tilde{M}a(T)$. The proposition follows. ■

The same result does not hold for the Slater solution. It is the case that $SL^* = \widetilde{SL}$ but this solution is strictly finer than the Uncovered Set, we shall prove that $SL^* \subset TC(UC)$.

<u>Lemma 5.1.10.</u> : The composition-extension, $\widetilde{SL}$, of the Slater solution is composition-consistent.

Proof :

For this proof, we need notations adapted to the study of two levels of decompositions. The bigger set X will be described as

$$X = \{(\alpha, \beta, \gamma) : \alpha \in \{1,..., k\}, \beta \in \{1,..., k(\alpha)\}, \text{ and } \gamma \in \{1,..., k(\alpha, \beta)\}\}.$$

A first level partition will be described as : $X = \cup_{\alpha \geq 1} X_\alpha$ with

$$X_\alpha = \{(\alpha, \beta, \gamma) : \beta \in \{1, ..., k(x)\}, \text{ and } \gamma \in \{1, ..., k(\alpha, \beta)\}\}.$$

Each X_α is partitioned : $X_\alpha = \cup_{\beta \geq 1} X_{\alpha,\beta}$ with

$$X_{\alpha,\beta} = \{(\alpha, \beta, \gamma) : \gamma \in \{1, ..., k(\alpha, \beta)\}\},$$

so that there is a second level partition of X :

$$X = \cup_{\alpha \geq 1,\, \beta \geq 1} X_{\alpha,\beta}.$$

If T is a tournament on X, T_α denotes the restriction of T to X_α and $T_{\alpha,\beta}$ denotes the restriction of T to $X_{\alpha,\beta}$ which is of course also the restriction of T_α to $X_{\alpha,\beta}$.

Suppose that each X_α is a component of T and that for each α, each $X_{\alpha,\beta}$ is a component of T_α. Then each $X_{\alpha,\beta}$ is also a component of T. We then denote by T^*_α, T^* and T^{**} the summaries defined by these decompositions, according to the formulae :

$\forall\, \alpha,\;\; T_\alpha = \Pi(T^*_\alpha; ..., T_{\alpha,\beta}, ...)$ $(\beta \in \{1, ..., k(\alpha)\})$,

$T = \Pi(T^*; ..., T_{\alpha,\beta}, ...)$ $(\alpha \in \{1, ..., k\}$ and $\beta \in \{1, ..., k(\alpha)\})$.

$T = \Pi(T^{**}; ..., T_\alpha, ...)$ $(\alpha \in \{1, ..., k\})$

These tournaments have orders

$$o(T_{\alpha,\beta}) = k(\alpha, \beta)$$

$$o(T_{\alpha}) = \sum_{\beta} k(\alpha, \beta)$$

$$o(T) = \sum_{\alpha} \sum_{\beta} k(\alpha, \beta)$$

$$o(T^*_{\alpha}) = k(\alpha)$$

$$o(T^*) = \sum_{\alpha} k(\alpha)$$

$$o(T^{**}) = k.$$

Here T^*_{α} is the restriction of T^* to the set $\{X_{\alpha,\beta} : \beta \in \{1, ..., k(\alpha)\}\}$, for instance one can write : $(\alpha, \beta_1)\ T^*_{\alpha}\ (\alpha, \beta_2)$, and it is straightforward that T^* can be decomposed according to T^{**} by :

$T^* = \Pi\,(T^{**} ; ..., T^*_{\alpha}, ...) \qquad\qquad (\alpha \in \{1, ..., k\}).$

With these notations let us prove the lemma. We consider $T^* = \Pi\,(T^{**} ; ..., T^*_{\alpha}, ...)$ and wish to prove the equality

$$\widetilde{SL}\,(T^*) = \cup\{\widetilde{SL}\,(T^*_{\alpha}) : \alpha \in \widetilde{SL}\,(T^{**})\}.$$

We first prove the inclusion of $\widetilde{SL}\,(T^*)$ in the later set and consider $(\alpha_0, \beta_0) \in \widetilde{SL}\,(T^*)$. There exists tournaments T, $T_{\alpha,\beta}$ such that $T = \Pi\,(T^* ; ..., T_{\alpha,\beta}, ...)$ and $X_{\alpha_0,\beta_0} \cap SL(T) \neq \emptyset$. Let $(\alpha_0, \beta_0, \gamma_0) \in X_{\alpha_0,\beta_0} \cap SL(T)$, there exists a Slater order U for T with $(\alpha_0, \beta_0, \gamma_0)$ on top. But X_{α_0} is a component of T, hence the restriction U_{α_0} of U to X_{α_0} can be taken as a Slater order for T_{α_0} , thus $(\alpha_0, \beta_0, \gamma_0) \in SL(T_{\alpha_0,\beta_0})$, which proves that $(\alpha_0, \beta_0) \in \widetilde{SL}\,(T^*_{\alpha_0})$. With the same Slater order, the decomposition $T = \Pi\,(T^{**} ; ..., T_{\alpha}, ...)$ proves that $\alpha_0 \in \widetilde{SL}\,(T^{**})$.

Conversely, let $\alpha_0 \in \widetilde{SL}\,(T^{**})$ and $(\alpha_0, \beta_0) \in \widetilde{SL}\,(T^*_{\alpha_0})$. We must find tournaments T and $T_{\alpha,\beta}$ such that $T = \Pi\,(T^*\,; \ldots, T_{\alpha,\beta}, \ldots)$ and $SL(T) \cap X_{\alpha_0,\beta_0} \neq \varnothing$.

Because $\alpha_0 \in \widetilde{SL}\,(T^{**})$, there exist a tournament T' and k tournament T'_α such that $T' = \Pi(T^{**}\,; \ldots, T'_\alpha, \ldots)$ and $SL(T') \cap X_{\alpha_0} \neq \varnothing$. By weak composition-consistency, this statement only depends on the orders, denoted n', $n'(\alpha)$, with $n' = \sum_\alpha n'(\alpha)$, of these tournaments. Moreover, if $n'(\alpha)$ (for $\alpha \in \{1, \ldots, k\}$) is a set of numbers such that $\alpha_0 \in \widetilde{SL}\,(T^{**})$ then it is easy to observe that, for any integer $\lambda > 0$, $\lambda\, n'(\alpha)$ leads to the same result. In particular, we can suppose that for every α, $n'(\alpha)$ is a multiple of the order $k(\alpha)$ of T^*_α : $n'(\alpha) = \mu(\alpha)\, n(\alpha)$, with $\mu(\alpha)$ a strictly positive integer.

Because $(\alpha_0, \beta_0) \in \widetilde{SL}\,(T^*_{\alpha_0})$ there exist tournaments T''_{α_0}, $T''_{\alpha_0,\beta}$ (for $\beta \in \{1, \ldots, k(\alpha_0)\}$) such that $T''_{\alpha_0} = \Pi(T^*_{\alpha_0}\,; \ldots, T''_{\alpha_0,\beta}, \ldots)$ and $SL(T''_{\alpha_0}) \cap X_{\alpha_0,\beta_0} \neq \varnothing$. Let $n''(\alpha_0, \beta)$ denote the order of $T''_{\alpha_0,\beta}$ and let $\lambda = n''(\alpha_0) = \sum_\beta n''(\alpha_0, \beta)$. For each α, let $n(\alpha) = \lambda\, n'(\alpha)$. We construct the tournament T in the following way :

- $T = \Pi\,(T^{**}\,; \ldots, T_\alpha\,, \ldots)$, where for each $\alpha \in \{1, \ldots, k\}$, $T_\alpha = \Pi\,(T^*_\alpha; \ldots, T_{\alpha,\beta}, \ldots)$ $\qquad (\beta \in \{1, \ldots, k(\alpha)\})$.
- For $\alpha \neq \alpha_0$ and for any $\beta \in \{1, \ldots, k(\alpha)\}$, $T_{\alpha,\beta}$ is any tournament of order $\lambda\, \mu(\alpha) = n(\alpha, \beta)$.
- For $\alpha = \alpha_0$ and for any $\beta \in \{1, \ldots, k(\alpha_0)\}$, $T_{\alpha_0,\beta}$ is any tournament of order $n''(\alpha_0, \beta)\, n'(\alpha_0) = n(\alpha_0, \beta)$.

Then the order of T_α for $\alpha \neq \alpha_0$ is $\lambda\, \mu(\alpha)\, k(\alpha) = \lambda\, n'(\alpha)$ and the order of T_{α_0} is $\sum_\beta n''(\alpha_0, \beta)\, n'(\alpha_0) = \lambda\, n'(\alpha_0)$. Hence the decomposition $T = \Pi\,(T^{**}\,; \ldots, T_\alpha\,, \ldots)$ shows that there exists a Slater order for T with the component T_{α_0} on top.

In the decomposition $T_{\alpha_0} = \Pi(T^*_{\alpha_0} ; ..., T_{\alpha_0,\beta}, ...)$ the order of $T_{\alpha_0,\beta}$ being $n''(\alpha_0 , \beta)\, n'(\alpha_0)$, there exists a Slater order for T_{α_0} with the component T_{α_0,β_0} on top. We deduce that there exists a Slater order for T with T_{α_0,β_0} on top, that is : $SL(T) \cap X_{\alpha_0,\beta_0} \neq \varnothing$. This proves that $(\alpha_0 , \beta_0) \in \widetilde{SL}\,(T^*)$ and the lemma. ■

Proposition 5.1.11.: The composition-consistent hull of the Slater solution is its composition-extension : $SL^* = \widetilde{SL}$.

Proof:

This is deduced from the preceding lemma by applying the proposition 2.5.4. ■

We now are in position to prove that $SL^* \neq UC$. More precisely, one has the following stronger result.

Proposition 5.1.12.: $SL^* \subset TC(UC)$ and $SL^* \neq TC(UC)$.

Proof:

We first prove the inclusion. Let $T \in \mathcal{T}(\{1, ..., k\})$ and $i \in \{1, ..., k\}$ such that $i \in SL^*(T)$. There exist tournaments $\widetilde{T}$, $\widetilde{T}_1$, ..., $\widetilde{T}_k$ such that $\widetilde{T} = \Pi\,(T; \widetilde{T}_1, ..., \widetilde{T}_k)$ with $SL(\widetilde{T}\,) \cap X_i \neq \varnothing$. By weak composition-consistency, we can choose the tournaments $\widetilde{T}_j$, $j \in \{1, ..., k\}$ to be linear orders. Denote x_{j_1} the Condorcet winner of $\widetilde{T}_j$. Then $x_{i_1} \in SL(\widetilde{T}\,)$ and, because $SL \subset TC(UC)$, $x_{i_1} \in TC(UC(\widetilde{T}\,))$. But, as one can easily check, $UC(\widetilde{T}\,) = \{\, x_{j_1} : j \in UC(T)\}$ and thus $TC(UC(\widetilde{T}\,)) = \{\, x_{j_1} : j \in TC(UC(T))\}$. Therefore, $i \in TC(UC(T))$. To prove that $SL^* \neq TC(UC)$ it is sufficient to prove that $TC(UC)$ is not composition consistent. To see this consider the tournament T of order 8 given in picture 5.2. For this tournament, x is a Condorcet loser in $UC(T)$; $x \in UC(T)$ but $x \notin TC(UC(T))$. Let a, b be two other points, and let $\widetilde{T} = \Pi(C_3\,; T, T_a, T_b)$ with C_3 the cyclone of order 3 and T_a and T_b the order-1 tournaments on $\{a\}$ and $\{b\}$. Then it is easy to see that $x \in TC(UC(\widetilde{T}))$, this proves that $TCoUC$ is not composition-consistent. ■

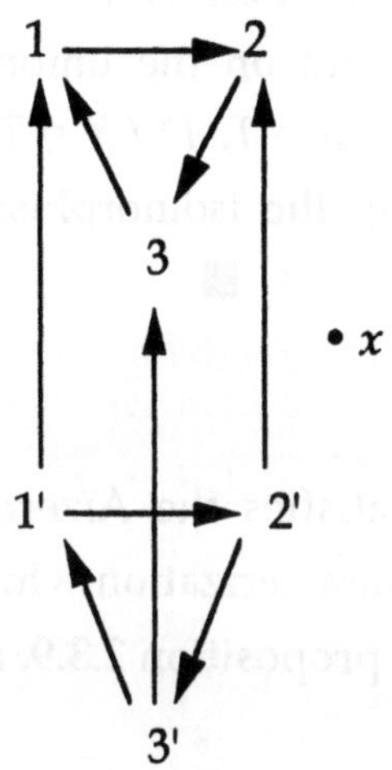

Figure 5.2.
(x is a Condorcet loser in $UC(T) = \{1, 2, 3, x\}$).

5.2. Iterations of the Uncovered Set

The Uncovered set is not idempotent, and it often selects many alternatives, so it is natural to consider the iterations UC^2, ..., UC^k, ..., UC^∞. These solutions are all different, as it can be easily deduced from the following proposition.

<u>Proposition 5.2.1.</u>: Let T be a tournament on a set X of at least 3 alternatives, with no Condorcet winner, there exists a tournament T^* of which T is a sub-tournament and such that $UC(T^*) = X$.

Proof:

The following construction is due to Moulin (1986). One considers a copy T' of T and on the union of the sets of vertices $X \cup X'$, T^* is defined by $T^*/X = T$, $T^*/X' = T$ and between X and X', $x'T^*x \Leftrightarrow x' = \varphi(x)$ (φ being the isomorphism mapping T onto T'). Then $UC(T^*) = X$. ■

The fact that UC satisfies the Aïzerman property allows to give for the limit UC^∞ a characterization which is slightly better than its definition. In effect, the proposition 2.3.9. applies and gives:

<u>Proposition 5.2.3.</u>: For all $x \in X - UC(X)$, $UC^\infty(X) = UC^\infty(X - \{x\})$).

Proof:

See 2.3.9. and 5.1.7., (v). ■

This characterization of UC^∞ permits the following algorithmic determination: «Find a covered alternative and remove it from the graph. Continue». The properties of UC^∞ are given in the next theorem. It should be pointed out that, even if it is clearly idempotent, and despite what could proposition 5.2.3. lead to think, UC^∞ does not verifies SSP. Even more surprisingly, UC^∞ is not monotonous. This is rather weird for a tournament solution, but shows that Monotonicity is not a trivial property.

<u>Theorem 5.2.4.</u>: The Iterated Uncovered set solution, UC^{∞},

(i) is not monotonous
(ii) is not independent of the losers
(iii) satisfies neither Aïzerman nor SSP
(iv) is idempotent
(v) is composition-consistent
(vi) is regular
(vii) is included in the Top-Cycle.

Proof :

(i) Let $T_1 \in \mathcal{T}(\{1, 2, 3, 1', 2', 3'\})$ be the tournament depicted in figure 5.3. (observe that this tournament is not the one of figure 5.1.).

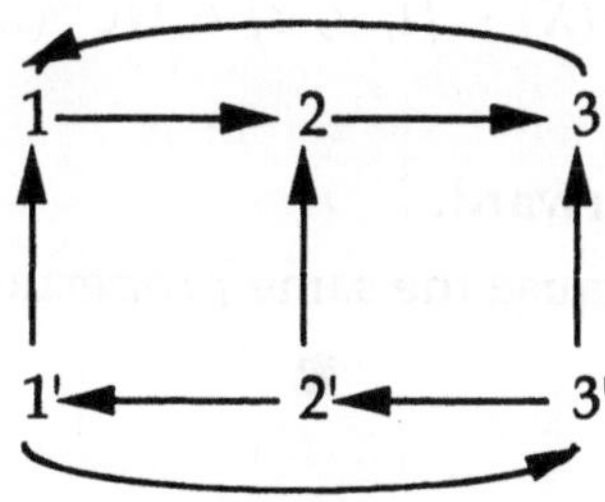

Figure 5.3.

It is easily checked that $UC(T_1) = X = UC^{\infty}(T_1)$, and in particular $1' \in UC^{\infty}(T_1)$. For the tournament $T_{1\,<1',\,2'>}$ one can (proposition 5.2.3.) successively eliminate the points 2' (covered by 1) then 3' (covered by 2) then 1' (covered by 3), so $UC^{\infty}(T_{1\,<1',\,2'>}) = \{1, 2, 3\}$, and $1' \notin UC^{\infty}(T_{1\,<1',\,2'>})$.

(ii) For the tournament $T_{1\,<1',\,2'>}$, reversing the arrow $1' \rightarrow 2'$ leads back to T_1 and gives the result.

(iii) Let $T_2 \in \mathcal{T}(\{1, 2, ..., 7\})$ be the tournament depicted in figure 5.4.

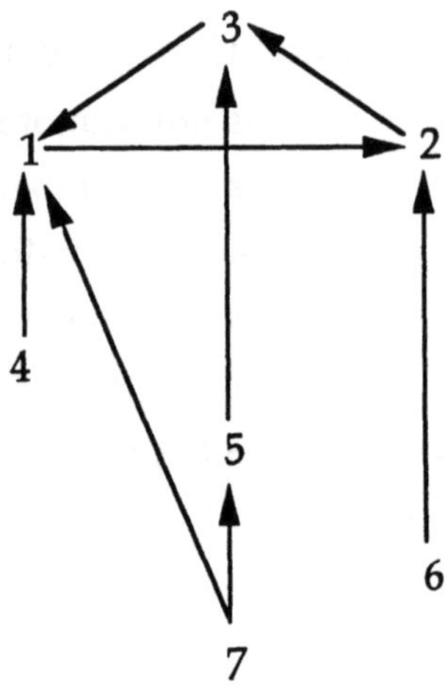

Figure 5.4.

In X, $4 \Rightarrow 7$ and $UC(X) = \{1, 2, 3, 4, 5, 6\}$, in $UC(X)$, $1 \Rightarrow 6$ and $UC^2(X) = \{1, 2, 3, 4, 5\}$, in $UC^2(X)$, $2 \Rightarrow 5$ and $UC^3(X) = \{1, 2, 3, 4\}$, in $UC^3(X)$, $3 \Rightarrow 4$ and $UC^4(X) = \{1, 2, 3\} = UC^\infty(X)$. One can check that $UC^\infty(X - \{4\}) = X - \{4\}$.

(iv) Straightforward.

(v), (vi), (vii) Because the same properties are true for *UC*.

■

The winners according to Copeland, Markov and Slater, although not covered, do not always belong to UC^2 and *a fortiori* to UC^∞:

Theorem 5.2.5.: (i) $C \not\subset UC^2$

(ii) $Ma \not\subset UC^2$

(iii) $SL \not\subset UC^2$

Proof:

(i) Here is an example. Let $T_1 \in \mathcal{T}(\{x_1, x_2, x_3, x_4, x_5\})$ be the tournament depicted in figure 5.5., for which $UC(T_1) = \{x_1, x_2, x_3, x_4\}$ and $UC^2(T_1)= \{x_1, x_2, x_3\}$.

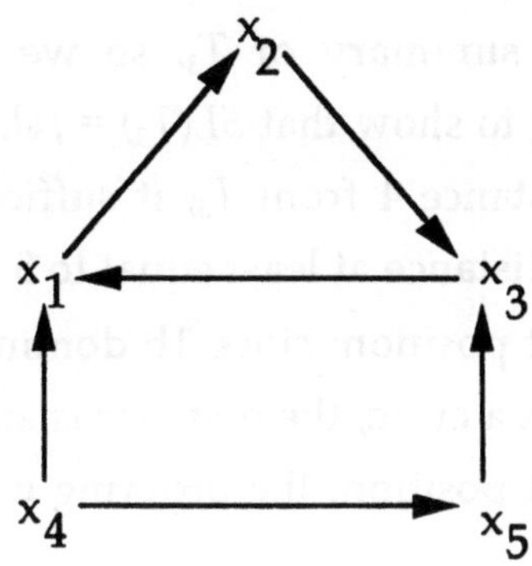

Figure 5.5.

Consider the tournament T_2 of order 11 obtained by composition from T_1 as summary and from the cyclical tournament C_3 of order 3 by : $T_2 = \Pi(T_1\,; C_3, \{x_2\}, \{x_3\}, \{x_4\}, C_3)$. One can check that the Copeland winner of T_2 is x_4 and that (by composition-consistency), $UC^2(T_2)$ is made of the five points of the components x_1, x_2 and x_3.

(ii) The same example works: computation of the Markov scores for the tournament T_2 gives the following result (up to a multiplicative constant): 37 for each point in the component x_1, 47 for x_2, 45 for x_3, 69 for x_4, and 9 for each point in the component x_5. As a consequence, $Ma(T_2) = \{x_4\}$.

(iii) Let $T_3 \in \mathcal{T}(\{1a, 1b, 2, 3, 4, 5a, 5b, 5c\})$ be the tournament depicted in figure 5.6.

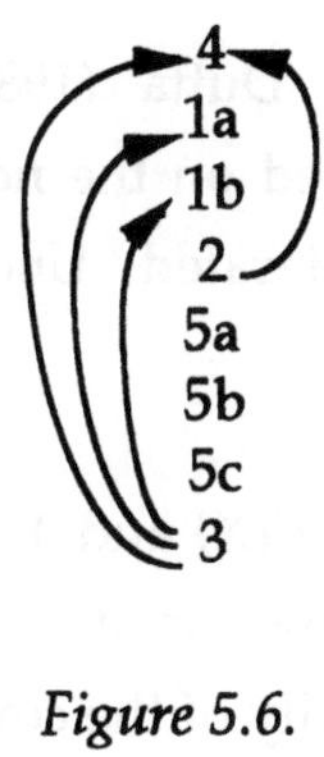

Figure 5.6.

Again, T_1 is a summary of T_3, so we find that $UC^2(T_3) = \{1a, 2, 3\}$. We are going to show that $SL(T_3) = \{4\}$. The figure 11 shows an ordering at the distance 4 from T_3, it suffices to prove that any other ordering is at a distance at least equal to 5 from T_3:

- if 1b is in last position: since 1b dominates 2, 5a, 5b and 5c and since 1a, 2, 3 forms a cycle, the distance is at least 5.
- If 1a is in last position: the ordering cannot be a Slater one because 1a covers 1b.
- Likewise, 2 and 4 cannot be in last position.
- If 5c is in last position: if such an ordering was a Slater one, then it would be possible to place 5a, 5b and 5c in the three last positions. Since they dominate 3 and since it remains the two cycles (4, 1a, 2) and (2, 3, 1b) we still have 5 reversed arrows.
- Since 5a and 5b cover 5c they cannot be in last position.
- If 3 is in last position, the reader will verify by the same methods that the Slater order of the restriction of T_3 to the set $\{$ 1a, 1b, 2, 4, 5a, 5b, 5c $\}$ is given by the figure 5.6, with 4 in top, which establishes the result. ■

5.3. Dutta's Minimal Covering Set

The solution proposed by Dutta (1988) is called the Minimal Covering set. It is again based on the notion of covering but it is a strict refinement of the iterated Uncovered set. Moreover, it possesses very nice properties.

Definition 5.3.1.: Let $T \in \mathcal{T}(X)$ and $Y \subset X$, Y is a *covering set* for T if for any $x \in X - Y$, $x \notin UC(Y \cup \{x\})$. Let $C(T)$ be the family of the covering sets for T.

A covering set is never empty. The family $C(T)$ is not empty, the whole set X is itself a covering set and other covering sets are given by the folowing proposition.

Proposition 5.3.2.: For any k, including $k=\infty$, $UC^k(T)$, is a covering set for T.

Proof :

First observe that, because the covering relation is transitive, when an alternative x is covered by an alternative, say y, in a tournament, one can always suppose that y itlself is not covered in this tournament. Let us prove that $UC(T) \in C(T)$. Let $x \in X - UC(T)$, x is covered in T by some $y \in UC(T)$; this means that y beats x and all the points of X that x beats, thus y *a fortiori* covers x in any subset of X containing x and y, and in particular y covers x in $UC(T) \cup \{x\}$. Hence $x \notin UC(UC(T) \cup \{x\})$, and we proved that $UC(T) \in C(T)$. Let us now prove the result for any finite k. Let $x \in X - UC^k(T)$, then for some $i<k$, $x \in UC^i(T)$ but $x \notin UC^{i+1}(T)$, there exists $y_{i+1} \in UC^{i+1}(T)$ covering x in $UC^i(T) \cup \{x\}$. Indeed, there exists a sequence x, y_{i+1}, y_{i+2}, ..., y_k such that $y_j \in UC^j(T)$ and either $y_j = y_{j+1}$ or y_{j+1} covers y_j in $UC^{j+1}(T) \cup \{ y_j \}$. Each y_j beats all the points that x beat in $UC^k(T)$, and x is beaten by y_k (if x were to beat y_k then any $y_j \neq y_k$ would beat y_k, impossible) thus y_k covers x is $UC^k(T) \cup \{x\}$, which proves that $UC^k(T) \in C(T)$. Since the result is true for any k it is also true for $UC^\infty(T)$. ■

The following proposition defines a tournament solution.

Proposition 5.3.3.: The family $C(T)$ admits a minimal (by inclusion) element called the *minimal covering set* of T and denoted by $MC(T)$.

Proof:

Since $\mathcal{C}(T)$ is non-empty and finite, it is sufficient to prove that $\mathcal{C}(T)$ is closed under intersection, then the minimal element of $\mathcal{C}(T)$ is simply the intersection of all the covering sets for T. The fact that $\mathcal{C}(T)$ is closed under intersection will be deduced from the following claim:

claim: Let X_1 and X_2 be two covering sets for $T \in \mathcal{T}(X)$. Let $x_0, x_1, ..., x_p$ be elements of X such that $x_0 \notin X_1$ and, for any $i \in \{1, ..., p\}$:

- if i is odd, $x_i \in X_1$ and x_i covers x_{i-1} in $X_1 \cup \{x_{i-1}\}$
- if i is even, $x_i \in X_2$ and x_i covers x_{i-1} in $X_2 \cup \{x_{i-1}\}$.

Then the restriction of T to $\{x_0, x_1, ..., x_p\}$ is transitive with $x_i \rightarrow x_{i-1}$.

This claim will be proven by induction on p. For $p = 1$ it is trivial. For $p > 1$, suppose the claim true up to $p - 1$. For a sequence $x_0, x_1, ..., x_p$ satisfying the above conditions, one can apply the induction hypothesis to the two sequences $x_0, ..., x_{p-1}$ and $x_1, ..., x_p$, hence we only have to show that $x_p \rightarrow x_0$. If p is odd, $p > 1$, $x_p \in X_1$ thus $x_p \neq x_0$, because x_1 covers x_0 in $X_1 \cup \{x_0\}$ and $x_p \rightarrow x_1$ we get : $x_p \rightarrow x_0$. If p is even, x_p covers x_{p-1} in $X_2 \cup \{x_{p-1}\}$, $x_{p-1} \rightarrow x_0$ and $x_0 \in X_2$ implies that $x_p \rightarrow x_0$. Hence the claim.

Let X_1 and X_2 be two covering sets. Suppose that $X_1 \cap X_2 = \emptyset$. Let x_0 be any point in X_2, because X_1 is covering there exists $x_1 \in X_1$ such that x_1 covers x_0 in $X_1 \cup \{x_0\}$, and it is possible to construct a sequence $x_0, ..., x_p$ for which the claim applies, for any p. But, obviously a finite tournament can not have transitive sub-tournaments of arbitrary length, hence $X_1 \cap X_2 \neq \emptyset$.

Let us prove now that $X_1 \cap X_2$ is a covering set. Let $x_0 \in X$, $x_0 \notin X_1 \cap X_2$. One may suppose $x_0 \in X_2$. Because X_1 is covering, there exists $x_1 \in X$, which covers x_0 in $X_1 \cup \{x_0\}$. If $x_1 \in X_1 \cap X_2$, x_1 also covers x_0 in $(X_1 \cap X_2) \cup \{x_0\}$; if $x_1 \notin X_2$, then there exists $x_2 \in X_2$ which covers x_1 in $X_2 \cup \{x_1\}$. Applying the claim, it is impossible to continue endlessly such a sequence, therefore, it must be the case that for some $i \geq 1$, $x_i \in X_1 \cap X_2$. We now prove that x_i covers x_0 in $(X_1 \cap X_2) \cup \{x_0\}$. From the claim we already know that $x_i \to x_0$. Let $y \in X_1 \cap X_2$ such that $x_0 \to y$. Because x_1 covers x_0 in $X_1 \cup \{x_0\}$ and $y \in X_1$, $x_1 \to y$; then because x_2 covers x_1 in $X_2 \cup \{x_1\}$ and $y_1 \in X_2$, $x_2 \to y$. Using i times the same argument one finally gets $x_i \to y$ and proves that x_0 is covered in $(X_1 \cap X_2) \cup \{x_0\}$. Hence $X_1 \cap X_2$ is a covering set. The theorem follows. ■

The original definition of a covering set by Bhaskar Dutta is different from definition 5.3.1. In the original definition, a covering set Y must satisfy the additional requirement that $UC(Y) = Y$. With this definition, the family of covering sets is no longer closed under intersection and the proofs are longer. Both definitions lead to the same concept of minimal set because of the following property.

<u>Proposition 5.3.4.</u>: $UC \circ MC = MC$

Proof :

Let $Y = MC(T)$ and $y \in Y$ such that $y \notin UC(Y)$. Denote $Z = Y - \{y\}$. Let us prove that Z is a covering set. Let $x \in X$, $x \notin Z$. If $x=y$, $Z \cup \{x\} = Y$, hence $x \notin UC(Z \cup \{x\})$. If $x \neq y$, $x \notin Y$ hence there exists $z \in Y$ such that z covers x in $Y \cup \{x\}$. If $z \neq y$ then $z \in Z$ and z also covers x in $Z \cup \{x\}$. If $z = y$ then $y \notin UC(Y)$ implies that there exists $z' \in Y$ such that z' covers y in Y: $z' \to y$ and $z' \to T^+(y) \cap Y$. Hence z' covers x in $Y \cup \{x\}$. But Z is a strict subset of Y, so Z being a covering set contradicts the minimality of Y. This proves that $UC(Y) = Y$ and the result. ■

Because $UC^\infty(T)$, is a covering set for T it is clear that the Minimal Covering set is finer than the Iterated Uncovered set. By mean of an example, B. Dutta proves that the former is sometimes strictly included in the latter. Comparing the two tournaments of order 6 depicted in figures 5.1. and 5.3., we can see that for both of them the Minimal Covering set is {1, 2, 3}, whereas the Uncovered set and its iterations are {1, 2, 3} in figure 5.1 and {1, 2, 3, 1', 2', 3'} in figure 5.3. Here are some properties of the Minimal Covering set, they should be contrasted with those of the Iterated Uncovered Set (theorem 5.2.4.).

<u>Theorem 5.3.5.</u>: The Minimal covering set solution, MC,

(i) is monotonous
(ii) is independent of the losers
(iii) satisfies SSP
(iv) is idempotent and satisfies Aïzerman
(v) is composition-consistent
(vi) is regular
(vii) is included in the Top-Cycle.

Proof :

(i) Let $x \in MC(T)$, and y such that $y \to x$. Denote $T' = T_{<x,\, y>}$ and $Y = MC(T')$. If $x \notin Y$ then $T/Y = T'/Y$, thus $UC(T/Y) = Y$ and for all $z \notin Y$, $z \notin UC(T/Y \cup \{z\})$, hence Y is covering for T, which contradicts $x \notin Y$.

(ii), (iii) and (iv): Straightforward in view of the definition.

(v) We do not reproduce the proof, which is easy although tedious (see Laffond, Lainé and Laslier 1996).

(vi) This fact will be proved later, with the help of the concept of tournament game (remark 6.3.4.).

(vii) Because $MC \subset UC$ (proposition 5.3.2.).

■

The relations with the preceding concepts are all deduced from the fact that MC is finer than UC^{∞} and we state them without proof (*cf.* 5.2.5 and 5.3.2)

Theorem 5.3.6.: (i) $MC \subset UC^{\infty}$ and $MC \neq UC^{\infty}$
(ii) $C \not\subset MC$
(iii) $Ma \not\subset MC$
(iv) $SL \not\subset MC$

B. Dutta has given an axiomatization of the Minimal Covering Set using the Strong Superset Property and a weakening of the expansion property called γ^*. The expansion property (5.1.5.) states that if $x \in X$ is chosen (according to S) from $Y \subset X$ and from $Y' \subset X$ then x is also chosen from $Y \cup Y'$. It can be equivalently stated: if $(Y_\alpha)_{\alpha \in A}$ is any family of subsets of X then, $x \in \bigcap_{\alpha \in A} S(Y_\alpha)$ implies $x \in S\left(\bigcup_{\alpha \in A} Y_\alpha\right)$. Dutta's axiom is the following:

Definition 5.3.7.: A solution S verifies the property γ^* if for any family $(Y_\alpha)_{\alpha \in A}$ of subsets of X, if $x \in \bigcap_{\alpha \in A} S(Y_\alpha)$ then $S\left(\bigcup_{\alpha \in A} Y_\alpha\right) \neq \bigcup_{\alpha \in A} Y_\alpha - \{x\}$.

Observe that in this axiom, one cannot replace the condition "For any class $(Y_\alpha)_{\alpha \in A}$ of subsets" by "For any two subsets Y_1 and Y_2".

Proposition 5.3.8.: The minimal covering set satisfies property γ^*.

Proof :

Let $T \in \mathcal{T}(X)$, with $X = \bigcup_{\alpha \in A} Y_\alpha$ and $x \in \bigcap_{\alpha \in A} (Y_\alpha)$. Suppose that $MC(T) = X - \{x\}$, this implies that x is covered in X by some y. Let α such that $y \in Y_\alpha$, y *a fortiori* covers x in Y_α, thus $x \notin UC(Y_\alpha)$ and thus $x \notin MC(Y_\alpha)$. ■

The property γ^{*} is sometimes satisfied by a tournament solution S for the mere reason that if S chooses all the alternatives but one, then this alternative is a Condorcet loser. This is the case of the Copeland and Markov solutions:

<u>Remark 5.3.9.</u>: If a tournament of order n admits exactly n-1 Copeland winners then it admits a Condorcet looser. Therefore, the Copeland solution trivially satisfies γ^{*}. The same holds for the Markov solution.

Proof :

Let T of order n be such that $C(T) = X - \{x\}$. Let $s(x)$ and $\bar{s}$ be the Copeland scores of x and of the other points. Then $s(x) \leq \bar{s} -1$ and the sum of all the scores is:

$$\frac{n(n-1)}{2} = s(x) + (n - 1)\ \bar{s}\,.$$

This implies that $\bar{s} \geq \frac{n-1}{2} + \frac{1}{n}$, and thus $\bar{s} \geq \frac{n}{2}$. It follows from the equation that $s(x)=0$.

For the Markov solution: Let T such that $Ma(T) = X - \{x\}$. Let $p(x)$ and $\bar{p}$ be the Markov scores of x and of the others points; $p(x) < \bar{p}$. Suppose that x is not a Condorcet loser, and take $y \in X$ such that $y \rightarrow x$ and z such that $x \rightarrow z$, then

$$s^{-}(y)\ \bar{p} = p(x) + [s(y) - 1]\ \bar{p} \qquad (1)$$

and

$$s^{-}(z)\ \bar{p} = s(z)\ \bar{p}\,. \qquad (2)$$

Thus all the elements of $T^{-}(x)$ have the same Copeland score, say $s(y)$ and all the elements of $T^{+}(x)$ have the same Copeland score, say $s(z)$. From equation (2) one finds : $s(z) = \frac{n-1}{2}$. The sum of the Copeland scores in T is:

$$\frac{n(n-1)}{2} = s^{-}(x)\, s(y) + s(x) + s(x)s(z)$$
$$= s^{-}(x)\, [s(y) - s(z)] + ns(z),$$

and thus one has: $s^{-}(x)\, [s(y) - s(z)] = 0$ and (if x is not a Condorcet loser) $s(y) = s(z)$. Then $s(y) = \frac{n-1}{2}$ and (1) implies $p(x) = \bar{p}$, a contradiction. ■

The property γ^{*} does not imply that if $S(T) = X - \{x\}$, x is a Condorcet loser, for instance, it is easy to find a tournament such that $MC(T) = X - \{x\}$. In this case, the point x must be covered and one must have: $UC(T) = X - \{x\}$. Indeed, property γ^{*} precisely implies that if $S(T) = X - \{x\}$, then x must be covered:

<u>Proposition 5.3.10.</u>: If S satisfies γ^{*} then there exists no tournament $T \in \mathcal{T}(X)$ such that $S(T) = X - \{x\}$ and $x \in UC(T)$.

Proof :

Suppose $S(T) = X - \{x\}$. For $y \in T^{-}(x)$, unless y covers x, there exists $z \in X$ such that $z \to y$ and $x \to z$. So if $x \in UC(T)$ one can consider a family $(Y_y)_{y \in T^{+}(x)}$ such that for each y, $Y_y = \{x, y, z\}$ with $x \to z$, $z \to y$ and $y \to z$. The restriction of T to Y_y is a 3-cycle and thus $S(Y_y) = Y_y$. Denote also $Y_x = T^{+}(x) \cup \{x\}$, then x is a Condorcet winner of T restricted to Y_x and $S(Y_x) = \{x\}$. Clearly $\cup \{Y_z, z \in T^{+}(x) \cup \{x\}\} = X$, therefore, γ^{*} implies that $S(T) \neq X - \{x\}$. ■

Call γ^{**} this weakening of γ^{*}:

<u>Definition 5.3.11.</u>: A solution S verifies the property γ^{**} if for any $T \in \mathcal{T}(X)$, if $S(T) = X - \{x\}$ then $x \notin UC(T)$.

The previous proposition just states that γ^{*} implies γ^{**}. It turns out that γ^{**} (rather than γ^{*}) is precisely the property used in Dutta's axiomatization of the Minimal Covering Set. One indeed has the following proposition.

Proposition 5.3.12.: Let S satisfy SSP and γ^{**}, then $S(T)$ is a covering set for T.

Proof :

For $y \notin S(T)$, denote $Y = S(T) \cup \{y\}$. By SSP, $S(Y) = Y - \{y\}$ thus, by γ^{**}, $y \notin UC(Y)$. Therefore, $S(T)$ is a covering set for T. ■

It is now easy to conclude:

Proposition 5.3.13.: The Minimal Covering Set is the finest tournament solution satisfying SSP and γ^{**}.

Proof:

We noticed (5.3.5. and 5.3.10.) that MC satisfies these two properties, the result follows from 5.3.12. and the definition of MC.

■

5.4. Weak Covering *à la* Laffond and Lainé

Recall that x covers y if x beats y and *all* the points that y beats. A natural weakening of the notion is then to say that x weakly covers y if x beats y and almost all the points that y beats. This idea is exploited by Laffond and Lainé (1994) under several variants. In the main one, "almost all" is taken to mean "all but maybe one".

Definition 5.4.1.: Let $T \in \mathcal{T}(X)$, $x, y \in X$. We say that x *weakly covers* y if xTy and $Card\ \{z \in X: yTz \text{ and } zTx\} \leq 1$

Clearly, if x covers y then x weakly covers y. Therefore, Gilbert Laffond and Jean Lainé defined a (sometimes strict) subset of the Uncovered Set.

<u>Definition 5.4.2.</u>: The *(-1)-Uncovered set* of T is the set of alternatives which are not weakly covered for T.

Unfortunately, it is possible that the (-1) Uncovered set be empty; for instance, if T is a 3-cycle ($x \to y$, $y \to z$, and $z \to x$) then the weak covering relation is also a cycle. The weak covering relation is not acyclic, so its set of maximal element is not the adequate concept for defining an associated tournament solution. Consider for instance a tournament of order 4 on $X = \{a, b, c, d\}$ with $\{a, b, c\}$ being a cyclic component and d a Condorcet loser. Then it is easy to compute the relation of weak covering for this tournament: $\{a, b, c\}$ forms a cycle for this relation and a, b, and c weakly covers d. In such an example, although every outcome is weakly covered, choice in X on the basis of weak covering should lead to the choice of $\{a, b, c\}$. Indeed, since the weak covering relation is not acyclic, one could consider its Top Set rather than its set of maximal elements (see 1.5.10. for the definition of the Top Set of a binary relation). Hence we state the following defintion.

<u>Definition 5.4.3.</u>: Given a tournament T, the Top Set of the weak covering relation of T is called the *Weak Uncovered Set* of T and is denoted by *WUC(T)*. Then *WUC* is a tournament solution.

The assertion that *WUC* is a tournament solution is true because any binary relation on a finite set admits at least one minimal retentive subset, retentive subsets being non-empty by definition (cf. 1.5.5. and 1.5.7.). Moreover, it is obvious that if x is a Condorcet winner for T then x weakly covers every other point. It is also clear that the Weak Uncovered Set is a refinement of the Uncovered Set. In order to see more precisely what happens when switching from the maximal elements of the weak covering relation ((-1)-Uncovered Set) to its Top Set (Weak Uncovered Set) we now study the cycles of the weak covering relation. The objective is a characterization of the Weak Uncovered set, this objective is achieved in proposition 5.4.12. The following proposition will prove very useful. Recall that $s(x)$ denotes the Copeland score of alternative x.

Proposition 5.4.4.: Let $T \in \mathcal{T}(X)$, $x, y \in X$. Suppose that x weakly covers y. Then $s(x) \geq s(y)$ and if $s(x) = s(y)$ there exists a unique $z \in X$ such that yTz and zTx, and there is no $z' \in X$ such that xTz' and $z'Tz$.

Proof :

Given two alternatives, x and y, the set X can be partitioned in 6 subsets: $\{x\}$, $\{y\}$, $T^-(x) \cap T^-(y)$, $T^-(x) \cap T^+(y)$, $T^+(x) \cap T^-(y)$, $T^+(x) \cap T^+(y)$. Definition 5.4.1. can be stated: xTy and $Card(T^-(x) \cap T^+(y)) \leq 1$. With these notations,

$$s(x) = Card(T^+(x) \cap T^+(y)) + Card(T^+(x) \cap T^-(y)) + 1$$
$$s(y) = Card(T^+(x) \cap T^+(y)) + Card(T^-(x) \cap T^+(y))$$
$$\leq Card(T^+(x) \cap T^+(y)) + 1.$$

Therefore $s(x) \geq s(y)$ and $s(x) = s(y)$ only if $Card(T^-(x) \cap T^+(y)) = 1$ and $Card(T^+(x) \cap T^-(y)) = 0$. ■

Definition 5.4.5.: Let $T \in \mathcal{T}(X)$, $x, y \in X$. We say that x *nearly covers* y for T if x weakly covers y and $s(x) = s(y)$.

Remark 5.4.6.: The four following statements are equivalent :

- x nearly covers y
- $T^+(x) = T^+(y) \cup \{y\} - \{z\}$
- $T^-(x) = T^-(y) \cup \{z\} - \{x\}$

- xTy, yTz, and zTx and $\{x, y\}$ is a component of $T/X\text{-}\{z\}$.

The "near-covering" relation is the right concept for studying cycles of the weak covering relation, as can be seen from the following proposition.

<u>Proposition 5.4.7.</u>: A sequence $(x_0, x_1, ..., x_n)$, $x_0 = x_n$, is a cycle for the weak covering relation for T if and only if it is a cycle for the "near-covering" relation for T.

Proof :

Suppose that $(x_0, x_1, ..., x_n)$ is a cycle for the weak covering relation. By proposition 5.4.4.,

$$s(x_0) \geq s(x_1) \geq ... \geq s(x_n),$$

thus $x_0 = x_n$ implies $s(x_0) = s(x_1) = ... = s(x_n)$. This in turn implies that $(x_0, x_1, ..., x_n)$ is a cycle for the "near-covering relation". The converse is trivial since x nearly covers y implies that x weakly covers y.

■

We now study the cycles of the "near-covering" relation.

<u>Proposition 5.4.8.</u>: (i) If x_1 and x_2 nearly covers y, then $x_1 = x_2$.
(ii) If x nearly covers y_1 and y_2, then $y_1 = y_2$.

Proof :

(i) If $x_1 \neq x_2$ one can suppose $x_1 \, T \, x_2$. But $T^+(x_1) \subset T^+(y) \cup \{y\}$. This contradicts $x_2 T y$.

(ii) Same argument: it is impossible that xTy, $y_1 T y_2$ and x nearly covers y_2. ■

Proposition 5.4.9.: Let $(x_0, x_1, ..., x_n)$, with $x_0 = x_n$ be a cycle of the relation "nearly covers for T". Then $\{x_1, ..., x_n\}$ is a component of T.

Proof:

Denote $Y = \{x_1, ..., x_n\}$ and take $z \in X\text{-}Y$. If for some $i \in \{1, ..., n\}$, $x_i T z$, then x_i nearly covers x_{i+1} and $z \neq x_{i+1}$ implies $x_{i+1} T z$. Repeating the argument implies that for all j, $x_j T z$. Therefore, z either beats all the elements of Y, either is beaten by all of them. This means that Y is a component of T. ■

If $Y \subset X$ forms a cycle for the "near-covering" relation, all the points in Y have the same score for T. Bus since, from the last proposition, they behave uniformly with respect to the points outside Y it is easy to deduce that they also have the same scores in the restricted tournament T/Y:

Remark 5.4.10.: Cycles of the "near covering" relation for T are regular components for T.

In fact, it is possible to completely characterize those regular components:

Proposition 5.4.11.: A subset Y forms a cycle of the "near-covering" relation for T if and only if Y is a component of T and T/Y is a cyclone of order greater than 1.

Proof :

Because T/Y is regular, it has odd order. Thus one can write $Y = \{x_0, x_1, ..., x_{2k}\}$ with, for any $i \in \{0, ..., 2k\}$, x_i nearly covers x_{i+1} for T (addition is modulo $2k+1$). For any i, proposition 5.4.9. implies that the unique z such that zTx_{i-1} and $x_i T z$ is in Y; this element can be denoted $x_{f(i)}$. Then one has :

$j \notin \{i\text{-}1, i, f(i)\} \Rightarrow (x_j\, T\, x_{i-1} \Leftrightarrow x_j\, T\, x_i)$.

Let us prove that f is a permutation of $\{0, ..., 2k\}$. Suppose for instance that for no i, $f(i) = 0$; then for all $i \in \{2, ..., 2k\}$, $x_0\, T\, x_{i-1} \Leftrightarrow x_0\, T\, x_i$. But $x_0\, T\, x_i$, therefore, one obtains that x_0 beats all the other points of Y. This contradicts regularity. This argument shows that f is a permutation of $\{0, ..., 2k\}$. We now prove that the successors of x_0 in Y are $x_1, x_2, ..., x_k$. Again this is sufficient for proving the proposition.

Let $f(j) = 0$ and $i \in \{1, ..., j\text{-}1\}$, then $0 \neq f(i)$ implies that $x_0\, T\, x_{i-1} \Leftrightarrow x_0\, T\, x_i$.

Let j be such that $f(j) = 0$ and let $i \in \{2, ..., j\text{-}1\}$. Then $j \neq i$ and f being a permutation implies that $0 \neq f(i)$ and therefore $x_0\, T\, x_{i-1} \Leftrightarrow x_0\, T\, x_i$. But $x_0 T\, x_1$, thus x_0 beats $x_1, x_2, ..., x_{j-1}$. On the other hand, if $i \in \{j+1, ..., 2k\}$, one finds that $x_i\, T\, x_0$. By definition of f, $x_j\, T\, x_0$, therefore x_0 is beaten by $x_j, x_{j+1}, ..., x_{2k}$. By regularity it must then be the case that $j = k+1$. The result follows. ■

We are now in position to characterize the elements of the Weak Uncovered Set.

<u>Proposition 5.4.12.</u>: Let $T \in \mathcal{T}(X)$, $x \in WUC(T)$ if and only if:

- either x is a maximal element of the weak covering relation,
- either x belongs to a component Y of T such that T/Y is a cyclone (of order greater than 1) and no point outside Y weakly covers x.

Proof:

Denote by R the weak covering relation for T. If x is a maximal element for R then $x \in WUC(T)$ because $Max(R) \subset TS(R) = WUC(T)$. If $x \in Y$ with Y a component of T and T/Y a cyclone, it is routine verification to check that Y forms a cycle for R. It is also easy to see that for any $x' \in Y$ and $z \in X\text{-}Y$, zRx' iff zRx ; therefore if no z

outside Y weakly covers x, Y is retentive for R. It follows that $Y \subset WUC(T)$ and $x \in WUC(T)$.

Conversely, let $x \in WUC(T)$. Let Y be the minimal retentive subset for R to which x belongs. If Y is a singleton then x is a maximal element for R because R is irreflexive: for no x, xRx. If Y is not a singleton, suppose that R/Y is acyclic, then it admits maximal element y (proposition 1.5.4.) and Y being retentive, y is also maximal with respect to X. This implies that $\{y\}$ is retentive and Y is not minimal retentive. Thus R/Y is cyclic. Let $Z \subset Y$ such that Z forms a cycle for R, by propositions 5.4.10. and 5.4.11., z is a component of T and T/Z is a cyclone. This implies that Z is retentive for R and thus $Z = Y$, the proof is complete. ■

We leave to the reader the exploration of the properties of the tournament solution WUC (for instance, WUC is monotonous and weakly composition-consistent). Observe that if one allows cloning of the alternatives, the definition of the weak covering as x beats y and *all except maybe one* alternatives beaten by y makes little sense, so that it is straightforward to see that the composition-consistent hull of WUC is the Uncovered Set. It has been remarked (proposition 5.4.4.) that if x weakly covers y then $s(x) \geq s(y)$ where s stands for the Copeland score. It follows that some Copeland winners must belong to the Weak Uncovered Set: for any T, $C(T) \cap WUC(T) \neq \emptyset$. Examples show that Copeland winners may not be in $WUC(T)$ and that elements of $WUC(T)$ may not be Copeland winners.

To conclude this section we come back to Laffond and Lainé's approach and mention one result which justifies consideration of the (-1)-Uncovered Set. We first introduce a generalization of the weak covering relation:

<u>Definition 5.4.13.</u>: Given $T \in \mathcal{T}(X)$, $\varepsilon \in \mathbb{N}$ and $x, y \in X$, we say that x *(-ε)-covers* y for T if xTy and

$$Card\{z \in X: yTz \text{ and } zTx\} \leq \varepsilon.$$

The (-ε)-Uncovered Set of T is the set of maximal elements for this relation and is denoted *(-ε)-UC(T)*.

For $\varepsilon = 0$ one obtains the usual covering relation and the Uncovered Set, which is non-empty. For $\varepsilon = 1$ one obtains the weak covering relation and the (-1)-Uncovered Set may be empty. The successive refinement of these sets are more likely to be empty, and for $\varepsilon = o(T) - 1$, the (-ε)-uncovered set is obviously empty.

<u>Theorem 5.4.14.</u>: Let $\varepsilon \in \mathbb{N}$ and $T \in \mathcal{T}(X)$ of order n. For $x \in X$ let $\Omega(\varepsilon, x)$ be the set of all subsets Y of X such that $x \in Y$ and $\#Y \geq n-\varepsilon$. Then $x \in (-\varepsilon)\text{-}UC(T)$ if and only if $x \in UC(T/Y)$ for all $Y \in \Omega(\varepsilon, x)$.

Proof :

Suppose that $x \notin (-\varepsilon)\text{-}UC(T)$. There exists $y \in X$ such that y (-ε)-covers x. Let $Z = \{z \in X: xTz \text{ and } zTy\}$, one has yTx and $\#Z \leq \varepsilon$. Let $Y = X - Z$, then $x, y \in Y$ and $\#Y \geq n - \varepsilon$, and y covers x in T/Y, hence we showed that there exists $Y \in \Omega(\varepsilon, x)$ such that $x \notin UC(T/Y)$. Conversely suppose that there exists $Y \in \Omega(\varepsilon, x)$ such that $x \notin UC(T/Y)$. There exists $y \in Y$ such that y covers x in the tournament T/Y. For this y one gets : yTx and for all $z \in Y$, $xTz \Rightarrow yTz$, therefore, $Card\{z \in X: xTz \text{ and } zTy\} \leq Card(X - Y) \leq \varepsilon$, and y (-ε)-covers x for T.

■

One interpretation for this result is the following: Take a set X of n players, the relations among these players being defined by tournament $T \in \mathcal{T}(X)$. Imagine that an actual competition is going to take place among these players but that some of the players will not participate in the actual tournament. Let Y be the set of players who actually show up. Suppose that player x is sure to come: $x \in Y$, and that the number of players who might stay home is limited: $Card\ Y \geq n - \varepsilon$. The question is: under which condition (on T) player x can be sure that she will actually be a winner (in T/Y). One answer is that, if "winners" are the uncovered players, then x is sure to win if and only if she belongs to the $(-\varepsilon)$-Uncovered Set of T.

5.5. Weak Covering *à la* Levchenkov

Here we define other weakenings of the covering relation, that is to say binary relations R on X such that xRy as soon as x covers y. Unlike the weakenings *à la* Laffond and Lainé, these relations are acyclic, so that they have maximal elements, which define refinements of the Uncovered set. But these relations are not always sub-relations of the tournament: it is possible that xRy even if yTx. A preliminary definition is useful.

Definition 5.5.1.: Given a tournament $T \in \mathcal{T}(X)$ and three alternatives x, y and z in X, we say that y is a *partner* of x against z if xTy and yTz.

This definition, as well as the other concepts of this section is due to Levchenkov (1995c). The notion of partner can be used to define the covering and weak covering relations. The same notion is also useful to state "two step principles" for the Uncovered set and the $(-\varepsilon)$-Uncovered sets. By refering to the definitions, the reader will easily check the statements in the following remarks.

<u>Remark 5.5.2.</u>: Let $T \in \mathcal{T}(X)$, $x, y \in X$, then:

(i) x covers y if
- y is beaten by x and
- y has no partner against x

(ii) x weakly covers y if
- y is beaten by x and
- y has no more than one partner against x

(iii) x $(-\varepsilon)$-covers y if
- y is beaten by x and
- y has no more than ε partners against x

<u>Remark 5.5.3.</u>: Let $T \in \mathcal{T}(X)$, $x \in X$, then:

(i) $x \in UC(T)$ if, for all $y \in X$:
- x beats y or
- x has at least one partner against y

(ii) x is in the (-1)-Uncovered set if, for all $y \in X$:
- x beats y or
- x has at least two partners against y

(iii) x is in the $(-\varepsilon)$-Uncovered set if, for all $y \in X$:
- x beats y or
- x has at least $(\varepsilon+1)$ partners against y

Levchenkov's definition can be stated in the following way.

<u>Definition 5.5.4.</u>: Let $T \in \mathcal{T}(X)$, $x, y \in X$. For any integer $q > 0$ we say that x *q-surpasses* y if
- y has no partner against x and
- either xTy or x has at least q partners against y.

The idea behind this definition is that one can say that x is stronger than y if x is beaten by y, provided that x has enough indirect support against y and that y has no indirect support against x.

<u>Proposition 5.5.5.</u>: Let $T \in \mathcal{T}(X)$, $x, y \in X$, $q \in \mathbb{N}$ with $q>0$, then x q-surpasses y if and only if

- either x covers y or
- yTx, $(\forall z \in X, yTz \Rightarrow xTz)$ and $Card\{z \in X\colon xTz \text{ and } zTy\} \geq q$

Proof :

It has already been noticed (remark 5.5.2.) that x covers y if and only if xTy and y has no partner against x. The proposition follows immediately. ■

<u>Proposition 5.5.6.</u>: Given $T \in \mathcal{T}(X)$, $x, y \in X$, for any two integers q and q' with $0 < q < q'$, if x q'-surpasses y then x q-surpasses y.

Proof :

Straightforward in view of the definition. ■

<u>Proposition 5.5.7.</u>: For $T \in \mathcal{T}(X)$ and $q > 0$, denote $SU_q(T)$ the set of maximal elements of the relation "q-surpasses" for T. Then :

$SU_1(T) \subset SU_2(T) \subset \ldots \subset SU_q(T) \subset \ldots \subset UC(T)$.

Proof :

From 5.5.5. and 5.5.6., for $0 < q < q'$, if x covers y then x q'-surpasses y and if x q'-surpasses y then x q-surpasses y. Therefore, if a point is maximal for the q-surpassing relation it is maximal for the q'-surpassing relation, and if it is maximal for the q'-surpassing relation, it is maximal for the covering relation. ■

The question is then the possible emptiness of these sets. It appears that for all $q \geq 2$, $SU_q(T)$ is non-empty:

<u>Proposition 5.5.8.</u>: Let $q \geq 2$, then the relation "q-surpasses" for T is acyclic and $SU_q(T)$ is non-empty.

Proof :

Observe that if x q-surpasses y then $s(x) \geq s(y) + q - 1$; thus for $q \geq 2$, $s(x) > s(y)$. Therefore, the q-surpassing relation is acyclic for $q \geq 2$. Thus its set of maximal elements is non-empty. ■

<u>Theorem 5.5.9.</u>: For any $q \geq 2$, SU_q is a tournament solution. For any tournament T, $C(T) \subset SU_q(T) \subset UC(T)$.

Proof :

Given $T \in \mathcal{T}(X)$, proposition 5.5.8. states that $SU_q(T)$ is a non-empty subset of X. If T has a Condorcet winner x_0, then $UC(T) = \{x_0\}$, therefore, by 5.5.7., $SU_q(T) = \{x_0\}$. It has been already noticed (in the proof of 5.5.6.) that if x q-surpasses y then $s(x) > s(y)$, hence if y is a Copeland winner, y is maximal for the q-surpassing relation.

■

Concerning the value $q = 1$, it is possible to make the same study that has been made in the previous section about the cycle of the weak covering relation: If x 1-surpasses y then either $s(x) > s(y)$ or $s(x) = s(y)$ and there exists a unique z such that $xTz\ Ty\ Tx$. But a cycle of the 1-surpassing relation is such that the Copeland score is constant along this cycle, therefore, one obtains (like in 5.4.14.) that cycles of the 1-surpassing relation are cyclone components of the tournament.

6 - Tournament Game

In this section we introduce another tournament solution called the Bipartisan set (Laffond, Laslier and Le Breton, 1991, 1993a). This solution is a refinement of the Minimal Covering set, although its definition does not explicitly use the notion of covering but the game-theoretical concept of Nash equilibrium. Apart from a solution concept, we also provide a scoring method (the "minimal gain score"). The games considered here have positive interpretations and several concepts of the previous chapters can be translated in the Game Theory framework.

6.1. Tournament Game in Pure Strategies

A two-player ("normal-form") game is a 4-uple: (X_1, X_2, g_1, g_2) where X_1 and X_2 are two sets and g_1 and g_2 are two applications from $X_1 \times X_2$ to $\mathbb{R}$. The set X_i is called the set of available strategies for player i and given $x_1 \in X_1$ and $x_2 \in X_2$, the number $g_i(x_1, x_2)$ is called the payoff for player i when player 1 plays x_1 and player 2 plays x_2.

Let $T \in \mathcal{T}(X)$, we associate to T a two-player zero-sum game in which each player has X as set of (pure) strategies. If $x_1 \in X$ is the strategy played by player 1 and $x_2 \in X$ is the strategy played by player 2, the payoff for player 1 is + 1 if $x_1 T x_2$, 0 if $x_1 = x_2$ and - 1 if $x_2 T x_1$.

Hence this game is of the form (X_1, X_2, g_1, g_2) with some particular features. First, it is a *finite* game: there is a finite number of strategies available to the players. Second, it is a *symmetric* game (with respect to the players); this means that $X_1 = X_2$ and, for all x_1 and x_2, $g_1(x_1, x_2) = g_2(x_2, x_1)$. One can speak of the payoff for strategy $x \in X$ against strategy y, without reference to the identity of the players. Symmetric games are, therefore, denoted (X, g), and in the case of a tournament game:

$$g(x, y) = g_1(x, y) = \begin{cases} +1 \text{ if } xTy \\ 0 \text{ if } x = y \\ -1 \text{ if } yTx \end{cases}$$

An additional important feature is that the game is *zero-sum*: for any x_1 and x_2, $g_1(x_1, x_2) + g_2(x_1, x_2) = 0$, that is: $g(x_1, x_2) + g(x_2, x_1) = 0$. This implies in particular that for all $x \in X$, $g(x, x) = 0$.

Definition 6.1.1.: A *tournament game* is a finite symmetric two-player game (X, g) such that, for all $x, y \in X$,

- $g(x, y) + g(y, x) = 0$ ("zero-sum").
- $x \neq y \Rightarrow g(x, y) \in \{-1, +1\}$.

Note that the matrix of the game is nothing but the comparison matrix which is analized from a geometrical point of view in Chapter 4. It is clear that any tournament defines a tournament game, and conversely, any tournament game (X, g) defines a tournament T on X : xTy if $g(x, y) = +1$. Given a tournament $T \in \mathcal{T}(X)$ we denote by $G(T) = (X, g^T)$ the associated tournament game. We now establish the connection between some concepts in Game Theory and concepts in Tournament Theory.

<u>Definition 6.1.2.</u>: In a game (X_1, X_2, g_1, g_2), a strategy $x_1 \in X_1$ is a *best response* to a strategy $x_2 \in X_2$ if

$$g_1(x_1, x_2) = Max_{y \in X_1} g_1(y, x_2).$$

<u>Proposition 6.1.3.</u>: In a tournament game $G(T)$, for any $y \in X$,
(i) if y is not a Condorcet winner of T then, any $x \in X$ such that xTy is a best response to y,
(ii) if y is a Condorcet winner of T then y is a best response to any other alternative and the unique best response to itself.

Proof :

If there exists $x_0 \in X$ such that x_0Ty, then, $g(x_0, y) = +1 = \underset{x}{Max}\, g(x', y)$ and for all x, x is a best response to y if and only if $g(x, y) = 1$, that is xTy. If y is a Condorcet winner, then for any $x \neq y$, $g(x, y) = -1$, hence $g(y, y) = 0$ implies that y is the unique best-response to itself ■

An important notion in Game Theory is the Nash equilibrium.

<u>Definition 6.1.4.</u>: A *pure Nash equilibrium* of the normal-form game (X_1, X_2, g_1, g_2) is a pair of strategies : $(x_1, x_2) \in X_1 \times X_2$ such that x_1 is a best response to x_2 and x_2 is a best response (for player 2) to x_1.

<u>Proposition 6.1.5.</u>: Let $G(T)$ be a tournament game. A pair (x, y) of strategies is a pure Nash equilibrium of $G(T)$ if and only if $x = y$ and x is a Condorcet winner of T.

Proof :

Let (x, y) be a Nash equilibrium of $G(T)$, if yTx, y is a better response than x to y : $g(x, y) = -1$ and $g(y, y) = 0$, therefore x is not a best response to y ; likewise, if xTy then y is not a best response to x, hence $x = y$. Suppose that y is not a Condorcet winner, then proposition 6.1.3. (i) indicates that y is not a best response to itself, hence (y, y) is not an equilibrium. Conversely, if y is a Condorcet winner, proposition 6.1.3. (ii) proves that (y, y) is an equilibrium.

■

Definition 6.1.6.: In a game (X_1, X_2, g_1, g_2), a strategy $x_1 \in X_1$ *dominates* another strategy $y_1 \in X_1$ if:

$$\forall x_2 \in X_2,\ g_1(x_1, x_2) \geq g_1(y_1, x_2),$$

with at least one strict inequality. If all these inequalities are strict, the domination is *strict*.

Proposition 6.1.7.: In a tournament game (with at least two alternatives), the strategy x dominates the strategy y if and only if x covers y.

Proof:

One just has to write definition 6.1.6. for a tournament game: for all $z \in X$, $g(x, z) \geq g(y, z)$, that is to say: if yTz then xTz and (for $y = z$), xTy. ■

The equivalence between domination in the game-theoretical sense and covering in the tournament terminology is noticed by McKelvey (1986). Unfortunately, it is often the case in the tournament literature that xTy is read "x dominates y". This is misleading and, since the meaning of the word "dominates" in Game Theory is well established, is avoided here.

<u>Remark 6.1.8.</u>: The Uncovered Set $UC(T)$ of a tournament T is the set of undominated strategies of the game $G(T)$. The iterated uncovered set, $UC^{\infty}(T)$ is the set of strategies which are not sequentially dominated.

These concepts are standard ones in Game Theory. For further details on them, the interested reader can refer, for instance, to Moulin (1981) or Myerson (1991).

Duggan and LeBreton (1996) have noticed that Dutta's Minimal Covering Set can also be given an interpretation as a game-theoretic concept. The concept to be used, the "Weak Saddle" of a game is due to Shapley (1964).

Notation: For a two-player game (X_1, X_2, g_1, g_2) and a non-empty subset Y_2 of X_2, let x_1 and x_i be in X_1. We write $x_1 >(Y_2)\, x_i$ if :

$$\forall y_2 \in Y_2,\ g_1(x_1, y_2) \geq g_1(x'_1, y_2)$$
$$\exists y_2 \in Y_2,\ g_1(x_1, y_2) > g_1(x'_1, y_2).$$

Similarly, we write $x_2 >(Y_1)\, x'_2$ if x_2 does better than x'_2 for player 2 when his opponent's strategies are restricted to $Y_1 \subset X_1$. If $x_1 >(Y_2)\, x'_1$, x_1 dominates x'_1 with respect to Y_2, the domination being "large with at least one strict inequality".

<u>Definition 6.1.9.</u>: Let (X_1, X_2, g_1, g_2) be a two-player game, $Y_1 \subset X_1$ and $Y_2 \subset X_2$.

(i) The pair (Y_1, Y_2) is a *Weak Generalized Saddle* if Y_1 and Y_2 are non empty and:

$$\forall x_1 \in X_1 - Y_1,\ \exists y_1 \in Y_1\colon y_1 >(Y_2)\, x_1$$
$$\forall x_2 \in X_2 - Y_2,\ \exists y_2 \in Y_2\colon y_2 >(Y_1)\, x_2.$$

(ii) If (Y_1, Y_2) is a Weak Generalized Saddle and if there exists no other Weak Generalized Saddle (Z_1, Z_2) such that $Z_1 \subset Y_1$ and $Z_2 \subset Y_2$ then we say that (Y_1, Y_2) is a *Weak Saddle.*

For finite zero-sum games, Weak Saddles always exist, but are not generally unique. But for tournament games one has:

Proposition 6.1.10.: Let T be a tournament, then $G(T)$ has a unique Weak Saddle: $(MC(T), MC(T))$.

Proof :

The equivalence between the definition of a Weak Saddle and the definition of the Minimal Covering set is clear, therefore the proposition is a by-product of the proof of the existence and unicity of the Minimal Covering set (proposition 5.3.3.). For more details, see Duggan and LeBreton (1996). ■

In the next section, we proceed to study the application to tournaments of another standard concept in Game Theory: the Mixed Nash equilibrium.

6.2. Tournament Game in Mixed Strategies

Given a game, its mixed strategies extension is another game, with the same players, in which players, instead of choosing the initial strategies, choose probability distributions over their strategy sets. Actual strategies are then assumed to be picked at random according to these distributions, independently from one player to another. The payoff in the mixed game is the mathematical expectation of the payoff in the initial game. The initial game is called the "pure" game. The formal definitions and notations are given now.

For a finite set X, the set of probability distributions on X is called the *simplex* on X and is denoted by Δ_X:

$$\Delta_X = \{p \in (\mathbb{R}_+)^X: \sum_{x \in X} p_x = 1\}.$$

For $p \in \Delta_X$, the *support* of p is: $Supp\ (p) = \{x \in X: p_x > 0\}$. We denote by δ_x the element of Δ_X with a unit mass on x.

Let (X_1, X_2, g_1, g_2) be a finite two-player game. For mixed strategies $p \in \Delta_{X_1}$ and $q \in \Delta_{X_2}$, we make a slight abuse of notation and write : $g_1(p, q) = \sum_{x_1 \in X_1} \sum_{x_2 \in X_2} p_{x_1} q_{x_2} g_1(x_1, x_2)$, and similarly for g_2. The associated *mixed game* is the two-player game $(\Delta_{X_1}, \Delta_{X_2}, g_1, g_2)$. If a game is symmetric and zero-sum, so is its associated mixed game, and one obtains for tournaments:

<u>Definition 6.2.1.</u> Let T be a tournament on X. The *mixed tournament game* associated to T is the two-player, zero-sum, symmetric game $J(T) = (\Delta_X, g^T)$ where, for any $p, q \in \Delta_X$, $g^T(p, q) = \sum_{x \in X} \sum_{y \in X} p_x q_y g^T(x, y)$.

General results for zero-sum symmetric games in mixed strategies apply for $J(T)$, and in particular the existence of a Nash equilibrium. Recall that the *value* of a two player zero-sum game $(X, Y, g, -g)$ is (whenever the equality holds) the number :

$$Max_{x \in X} \{Min_{y \in Y}\, g(x,y)\} = Min_{y \in Y} \{Max_{x \in X}\, g(x,y)\}.$$

<u>Proposition 6.2.2.</u>: The game $J(T)$ has an equilibrium. Its value is zero and it has a symmetric equilibrium.

Proof :

The set X being finite this proposition is deduced from the elementary theory of zero-sum games. See for instance Owen (1982).

■

Let $E(T) \subset \Delta_X \times \Delta_X$ be the set of equilibria of $J(T)$, it is possible to directly characterize the symmetric equilibria. For $A \subset X$, let us denote $p[A] = \sum_{x \in A} p_x$.

<u>Proposition 6.2.3.</u>: Let $p \in \Delta_X$, then $(p, p) \in E(T)$ if and only if for all $x \in X$:

(i) $x \in Supp(p) \Rightarrow p[T^+(x)] = p[T^-(x)]$

(ii) $x \notin Supp(p) \Rightarrow p[T^-(x)] \geq 1/2$.

Proof:

Let $(p, p) \in E(T)$ and $x \in X$, let $\delta_x \in \Delta_x$ the probability of support $\{x\}$; the payoff for playing p against δ_x is positive and can be written $p[T^-(x)] - p[T^+(x)]$, thus one has :

$$\forall x \in X, p[T^-(x)] \geq p[T^+(x)]. \tag{1}$$

If $p_x = 0$, $p[T^+(x)] + p[T^-(x)] = 1$, which gives (ii). To obtain (i), let us write that the value of the game is zero :

$$0 = \sum_{x \in X} p_x \left(\sum_{y \in T^-(x)} p_y - \sum_{y \in T^+(x)} p_y\right)$$

that is :

$$0 = \sum_{x \in Supp(p)} p_x \{p[T^-(x)] - p[T^+(x)]\}, \tag{2}$$

and (i) is deduced from (1) and (2). Conversely, for all $q \in \Delta_X$, the payoff for playing q against p is $\sum_{x \in X} q_x \{p[T^+(x)] - p[T^-(x)]\}$, negative if (i) and (ii) are true; this proves that (p, p) is an equilibrium. ■

The game and its equilibria can be rewritten in the language of linear algebra: let M be the tournament matrix, whose entry m_{ij} is 1 if iTj and 0 if not, the matrix of the game is the comparison matrix, N, whose entries are $n_{ij} = 1$ if iTj, $n_{ij} = 0$ if $i = j$ and $n_{ij} = -1$ if jTi is $N = M - M^T$. We are going to prove that the game $J(T)$ has in fact a unique equilibrium. This result can be obtained by pure linear algebra methods (Fisher and Ryan, 1992, 1995a, b, c), but we shall use another method, which allows to make the link between the equilibrium of the game and the tournaments which are summaries of regular tournaments.

<u>Proposition 6.2.4.</u>: Let $T \in \mathcal{T}(X)$ and $p \in \Delta_X$. If $(p, p) \in E(T)$ then $T/Supp(p)$ is the summary of a tournament $T^* = \Pi\,(T/Supp(p); \ldots T^*_x \ldots)$ such that:

- for all $x \in Supp(p)$, $p_x = \dfrac{o(T^*_x)}{o(T^*)}$ and $o(T_x)$ is odd
- $\# Supp(p)$ is odd
- T^* is regular (and $o(T^*)$ is odd)

Proof:

The solutions of the system (i) of proposition 6.2.3. are solutions of a system of linear *equations* with integers coefficients, hence they are rational numbers. Thus there exists an integer l such that for all $x \in X$, $q_x = l\, p_x$ is an integer, moreover it is possible to choose l such that at least one of the integers q_x is odd. We still have for all $x \in Supp(p)$:

$$\sum_{y \in T^-(x)} p_y = \sum_{y \in T^+(x)} p_y.$$

But $l = \sum_{x \in X} q_x$, so $l - q_x = 2 \sum_{y \in T^+(x)} q_y$ shows that q_x has the same parity as l, for all $x \in Supp(p)$. One of the integers q_x being odd, l is odd and all the q_x are odd. This prove firstly that $\#Supp(p)$ is odd and secondly it allows to build the tournament T^* as the product of the tournament $T/Supp(p)$ by regular tournaments T^*_x of order q_x, for instance the cyclical tournaments C_{q_x}. It is easily deduced from this construction that T^* is regular of order l. ■

The previous proposition admits a converse :

<u>Proposition 6.2.5.</u>: Let $T \in \mathcal{T}(X)$ be a tournament summary of a regular tournament $T^* \in \mathcal{T}(X^*)$. Let $n = o(T)$, for $x \in X$ let $q_x = \#x$ be the cardinal of the component x of T^* and let $p_x = q_x/n$. Then $(p, p) \in E(\tilde{T})$.

Proof :

Let $x \in X$ be a component of T^*, from proposition 1.4.6., T^*/x is regular. Regularity implies :

$$(q_x - 1)/2 + \sum_{y \in T^+(x)} q_y = (q_x - 1)/2 + \sum_{y \in T^-(x)} q_y$$

which gives equation (i) of proposition 6.2.3. for T and the result.

■

These two propositions give a complete characterization of the tournaments for which there exists a symmetric equilibrium of the associated game giving strictly positive weight to all the edges. We can now proceed to prove unicity of the (mixed) Nash equilibrium for tournament games.

<u>Theorem 6.2.6.</u>: The game $J(T)$ has a unique Nash equilibrium and this equilibrium is symmetric.

Proof :

The proof goes in two steps.

First step : Let p and p' be such that $(p, p) \in E(T)$ and $(p', p') \in E(T)$, with $Supp(p) = Supp(p') = Y$, then $p = p'$.

From proposition 6.2.4. there exists a regular tournament $T^* \in \mathcal{T}(Y^*)$ such that T/Y is a summary of T^*, for $y \in Y$, the cardinal of the component y of T^* being odd. For $z \in Y^*$ let $y(z)$ be the component to which z belongs and let $q(z) = \dfrac{p'(y(z))}{\# y(z)}$. By construction $(q, q) \in E(T^*)$, thus there exists another regular tournament T^{**} of which T^* is a summary. From proposition 1.4.6., all the components of T^{**} have the same order, thus $q(z)$ does not depend of z, thus $p'(y)$ is proportional to $\#y$, which is sufficient to prove that $p' = p$.

Second step : There is one and only one p such that $(p, p) \in E(T)$.

In effect, the only possibility of non-uniqueness remaining after the first step is the case of two equilibria (p, p) and (p', p') with supports Y and Y' different. But $E(T)$ being convex, for all $\lambda \in]0, 1[$, $(\lambda p + (1 - \lambda) p', \lambda p + (1 - \lambda) p') \in E(T)$, and the support of $\lambda p + (1 - \lambda)p'$ is $Y \cup Y'$. According to the first step, there cannot be several symmetric equilibria with the same support $Y \cup Y'$, thus $p = p'$ and $Y = Y'$.

End of the proof : Let (p, q) be a Nash equilibrium, then (p, p) and (q, q) are also equilibria; the second step implies that $p = q$ and the theorem. ■

It is possible to say a little more, from a game-theoretic point of view, about the unique mixed Nash equilibria of a tournament game.

<u>Definition 6.2.7.</u>: Let (p, q) be a mixed Nash equilibria of the game (X_1, X_2, g_1, g_2) ; we say that (p, q) is a *strict* mixed Nash equilibria if :

$$\forall x_1 \in X_1,\ x_1 \notin Supp(p) \Rightarrow g_1(x_1, q) < g_1(p, q)$$
$$\forall x_2 \in X_2,\ x_2 \notin Supp(p) \Rightarrow g_2(p, x_2) < g_2(p, q).$$

At a mixed Nash equilibrium (p, q), all the strategies $x_1 \in Supp(p)$ which are played with positive probability yield the same payoff for a player against the opponent's equilibrium strategy:

$$\forall x_1 \in X, x_1 \in Supp(p) \Rightarrow g_1(x_1, q) = g_1(p, q)$$

and the strategies which are not played yield a payoff which is lower or equal:

$$\forall x_1 \in X, x_1 \notin Supp(p) \Rightarrow g_1(x_1, q) \leq g_1(p, q).$$

At a strict mixed Nash equilibrium, deviating from the support of the equilibrium induces strict losses.

Proposition 6.2.8.: The mixed Nash equilibrium of a tournament is strict.

Proof:

Let (p, p) be the equilibrium of $J(T)$, we only need to show that if $x \notin Supp(p)$, $g(x, p) \neq 0$. For any x, $g(x, p) \neq 0$ is equivalent to $p[T^+(x)] \neq ½$, but, from proposition 6.2.4., $p[T^+(x)]$ can be written as: $p[T^+(x)] = \sum_{y \in T^+(x) \cap Supp(p)} o(T^*_y)/o(T^*)$ with $o(T^*)$ odd. Therefore, this ratio can not equal ½ and $g(x, p) \neq 0$. ■

It should be remarked that the existence of decomposable regular tournaments (Theorem 1.4.7.) proves that the equilibrium strategy p is not necessarily uniform. This concludes the study of the equilibrium of the game $J(T)$, and suggests a new solution concept consisting in choosing the support of p. A refinement of this solution could be to choose the alternatives which are played at equilibrium with the highest probability. But, in view of proposition 6.2.4., the interpretation of equilibrium probabilities as indicators of "strength" in the tournament is misleading. For instance, if one reinforces a point x played at equilibrium with a positive probability p_x it remains true that x is played with a positive probability p'_x (see further theorem 6.3.2.), but it may be the case that p'_x is in fact *smaller* than p_x. In section 3 of this chapter we shall introduce, using the tournament game, another method for ranking the alternatives.

Example 6.2.9.: Here is an example of the "paradoxical" behavior of equilibrium weights.

Let $X = \{1, 2, 3, 4, 5, 6, 7\}$ and let T_1 and T_2 be the two tournaments depicted in figure 6.1. In order for the picture to be more clear, some arrows are summarized by double arrows. The complete description of T_1 is :

$1 \rightarrow 2\,3\,4$

$2 \rightarrow 3\,5\,6\,7$

$3 \rightarrow 4\,5\,6\,7$

$4 \rightarrow 2\,6\,7$

$5 \rightarrow 1\,4\,6$

$6 \rightarrow 1\,7$

$7 \rightarrow 1\,5$

And the equilibrium weights for T_1 are:

$1 : 7/25$

$2 : 1/5$

$3 : 1/25$

$4 : 3/25$

$5 : 3/25$

$6 : 1/25$

$7 : 1/5$

The tournament T_2 is equal to T_1 $_{<2,\ 7>}$, that is T_2 is identical to T_1 except that alternative 7 has been strengthened against alternative 2: $2T_17$ but $7T_22$. The equilibrium weights for T_2 are also all positive :

$1 : 9/35$

$2: 1/7$

$3 : 1/5$

4 : 1/35
5 : 1/35
6 : 1/5
7 : 1/7

But the weight of alternative 7 is 1/5 in T_1 and only 1/7 in T_2. (These two tournaments, with their associated equilibrium weights, are taken from Fisher and Ryan (1995a), who provide a complete list of the tournaments of order seven such that all the alternatives have positive weights.)

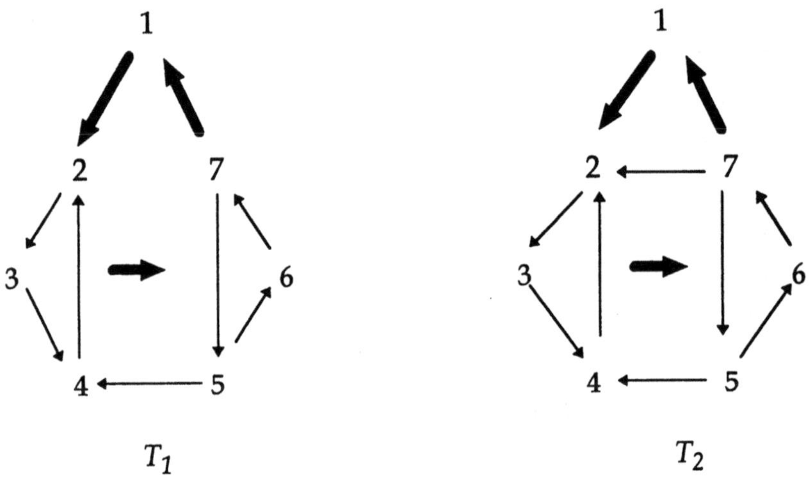

Figure 6.1.

6.3. Properties of the Bipartisan Set

<u>Definition 6.3.1.</u> : Let T be a tournament, we denote by $BP(T)$, and we call *Bipartisan set* of T the support of p, where (p, p) is the unique equilibrium of the game $J(T)$.

From the analysis of the previous paragraphs, BP is a tournament solution, $BP(T)$ always has an odd number of elements, and is a singleton if and only if T has a Condorcet winner. This tournament solution has many properties in common with the Minimal Covering set (compare the following theorem with theorem 5.3.5.).

Theorem 6.3.2. : The Bipartisan solution, BP,

(i) is monotonous
(ii) is independent of the losers
(iii) satisfies SSP
(iv) is idempotent and satisfies the Aïzerman property
(v) is composition-consistent
(vi) is regular
(vii) is included in the Top-cycle.

Proof :

(ii), (iii) and (iv): These points are immediately deduced from proposition 6.2.3.

(i) Let $x \in BP(T)$ and $y \in T^-(x)$. Let $T' = T_{<x, y>}$ and let p be the equilibrium strategy for T. We have to prove $x \in BP(T')$. If $y \notin BP(T)$ the relations of proposition 6.2.3. still hold in T', thus $E(T') = \{(p, p)\}$ and $BP(T') = BP(T)$, in particular, $x \in BP(T')$. If $y \in BP(T)$ then two exclusive cases must be distinguished :

a) $y \in BP(T')$. If $x \notin BP(T')$ then, according to the first part of the proof applied to tournament T', one should have, because (obviously) $T'_{<y, x>} = T$, $BP(T') = BP(T)$, which is impossible. Hence $x \in BP(T')$.

b) $y \notin BP(T')$. If $x \notin BP(T')$ then $BP(T) = BP(T')$, from the property of independence of the losers. Again, $x \in BP(T')$.

(v) Let $T \in \mathcal{T}(X)$ be a composed tournament and $U \in \mathcal{T}(A)$ be a summary of T. For $x \in X$ let us denote by $a(x)$ its component, and let us denote by $p = (p_a)_{a \in A}$ the equilibrium strategy for U and by $q^a = (q^a{}_x)_{x \in A}$ the equilibrium strategy for T/a. For $x \in X$, let $r_x = p_a(x)$

$q_x^{a(x)}$, it is easily checked that $r \in \Delta_X$, we shall show that r is the equilibrium strategy for T. It is sufficient to verify the conditions of proposition 6.2.3. For all $x \in X$, $x \in Supp(r)$ if and only if $x \in Supp\,(q^{a(x)})$ and $a(x) \in Supp(p)$, moreover,

$$\sum_{y\in T^-(x)} r(x) = \sum_{b\in U^-(a(x))} p_b + p_{a(x)} \sum_{y\in T^-(x)\,\cap\, a(x)} q^{a(x)}{}_y$$

and

$$\sum_{y\in T^+(x)} r(x) = \sum_{b\in U^+(a(x))} p_b + p_{a(x)} \sum_{y\in T^+(x)\,\cap\, a(x)} q^{a(x)}{}_y.$$

But since (p, p) is the equilibrium for U,

$$\sum_{b\in U^+(a(x))} p_b \leq \sum_{b\in U^-(a(x))} p_b$$

with equality if $p_{a(x)} \neq 0$, and similarly

$$\sum_{y\in T^+(x)\,\cap\, a(x)} q^{a(x)}{}_y \leq \sum_{y\in T^-(x)\,\cap\, a(x)} q^{a(x)}{}_y$$

with equality if $q^{a(x)}{}_y \neq 0$. We deduce that :

$$\sum_{y\in T^+(x)} r(x) \leq \sum_{y\in T^-(x)} r(x)$$

with equality if $r(x) \neq 0$, hence the result.

(vi) Straightforward. Observe that, for a regular tournament, the equilibrium strategy is the uniform distribution.

(vii) is deduced from (v).

■

A tournament $T \in \mathcal{T}(X)$ satisfies $BP(T) = X$ if and only if it is the summary of a regular tournament. Fisher et Ryan (1995a) call "positive" the tournaments for which the optimal strategy p is a strictly positive vector. Therefore, the results above characterize the positive tournaments.

As for Minimal Covering set, the properties satisfied by the Bipartisan set are intuitively very attractive. One question naturally arises : are the two solutions identical? The answer is no, indeed, BP is a refinement of MC, sometimes strict :

Theorem 6.3.3. : $BP \subset MC$ and $BP \neq MC$.

Proof :

First, we prove the inclusion $BP \subset MC$. Let $T \in \mathcal{T}(X)$, and let p be the equilibrium strategy for $T/MC(T)$. Denote again by p (slight abuse of notation) the probability distribution extended to X by letting $p(x) = 0$ if $x \notin MC(T)$. We prove that p is the equilibrium strategy for T. To do so, it is sufficient to prove that for all $x \notin MC(T)$, $p[T^+(x)] \leq p[T^-(x)]$, that is, since p is zero outside $MC(T)$: $p([T^+(x) \cap MC(T)] \leq p[T^-(X) \cap MC(T)]$. By definition of $MC(T)$ there exists an alternative $y \in MC(T)$ which covers x in $MC(T) \cup \{x\}$, that is $y \in T^-(x)$ and $T^+(x) \cap MC(T) \subset T^+(y) \cap MC(T)$. We deduce that $p[T^+(x) \cap MC(T)] \leq p[T^+(y) \cap MC(T)]$. But the definition of p gives $p[T^+(y) \cap MC(T)] \leq p[T^-(y) \cap MC(T)]$ and thus $p[T^+(x) \cap MC(T)] \leq p[T^+(y) \cap MC(T)] \leq p[T^-(y) \cap MC(T)] \leq P[T^-(x) \cap MC(T)]$ and the result.

To establish that BP is strictly more selective than MC, the counter-example (of order 6) is the one of figure 6.2. One verifies at hand that $MC(T) = \{1, 2, 3, 4, 5, 6\}$, and the result is deduced from the observation that $BP(T)$ has an odd number of points (computation gives here $BP(T) = \{1, 2, 3, 4, 6\}$, with weights $1, 1, 3, 3,$ and 1).

■

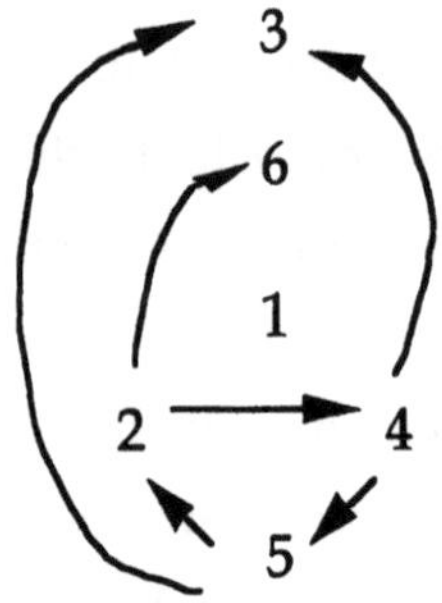

Figure 6.2.

<u>Remark 6.3.4.</u>: The inclusion $BP \subset MC$ allows to deduce the regularity of MC, announced in theorem 5.3.3. Also it allows to make precise the relations with the other solution concepts :

<u>Corollary 6.3.5.</u>:

(i) $BP \subset MC \subset UC^{\infty} \subset UC^{k} \subset UC$

(ii) $C \varnothing BP$

(iii) $Ma \varnothing BP$

(iv) $SL \varnothing BP$.

The next proposition indicates that, although the Bipartisan set is a refinement of the Minimal Covering set, it fails to define *unique* winners in all non-trivial cases where there is no Condorcet winner. By contrast, scoring methods often specify unique winners. In the absence of a Condorcet winner, the Bipartisan set contains at least three alternatives. Note that such is not the case for the two notions of Weak Uncovered set presented in section 4 and 5 of chapter 6 (see the example in the annex).

<u>Proposition 6.3.6.</u>: For the Bipartisan set, the Minimal Covering set, the Iterated Uncovered sets, the Uncovered set and the Top-Cycle, the winner is unique if and only if it is a Condorcet winner, and if there is no Condorcet winner then the set of winners contains at least three alternatives.

Proof :

If the support of the Nash equilibrium in mixed strategies is a singleton it is clear that this alternative defines a Nash equilibrium for the game in pure strategies. Therefore the proposition for the Bipartisan set is a consequence of the remark that the Bipartisan set always contain an odd number of alternatives. The proposition then also holds for the other mentioned solutions because they are coarser than BP. ■

We end this section by an axiomatization of the Bipartisan set as a tournament solution.

<u>Definition 6.3.7.</u>: Let $T \in \mathcal{T}(X)$, $x \in X$ and $Y \subset X$. The Copeland score of x with respect to Y is :

$$s(x, Y) = \#\{y \in Y : xTy\}.$$

<u>Definition 6.3.8.</u>: A subset $Y \neq \emptyset$ of X is *regular* for $T \in \mathcal{T}(X)$ if T/Y is regular, moreover, Y is *Copeland-dominant* if :

$$\forall x \in X\text{-}Y,\ s(x, Y) < \frac{\#Y}{2}.$$

<u>Definition 6.3.9.</u>: A tournament solution S is *Copeland-dominant* if, for any tournament T, if there exists a regular and Copeland dominant subset Y for T then $Y \subset S(T)$.

<u>Theorem 6.3.10.</u>: The Bipartisan tournament solution is the finest tournament solution S such that

- S is composition-consistent
- S is Copeland-dominant.

Proof:

It has already been noticed that BP is composition-consistent. Let Y be a regular and Copeland-dominant subset for $T \in \mathcal{T}(X)$. Define p on X by $p_x = 0$ if $x \in X-Y$ and $p_y = 1/(\#Y)$ if $y \in Y$. Then it is easily checked that p is the equilibrium probability for T, hence $BP(T) = Y$. This proves that BP satisfies the two axioms of the proposition.

Let S be a tournament solution and $T \in \mathcal{T}(X)$. Denote by p the equilibrium probability for T and let $Y = BP(T) = Supp(p)$. From proposition 6.2.4., T/Y is the summary of a regular tournament on a set Y^*. Denote $T^*/Y^* \in \mathcal{T}(Y^*)$ this tournament, Y is a partition of Y^* and for $y \in Y$, p_y is the proportion of the component y in Y^* : $p_y = \dfrac{\#y}{\#Y^*}$. Let $X^* = Y^* \cup (X - Y)$ and construct the tournament T^* on X^* such that $T^*/(X - Y) = T/(X - Y)$ and, for $y^* \in Y^*$ and $x \in X - Y$, $y^* T^* x$ if yTx where y is the component to which y^* belongs. Then, in T^*, Y^* is regular, moreover, if $x \in X - Y$, the Copeland score of x with respect to Y^* is:

$$s(x, Y^*) = \sum_{y \in Y \cap T^+(X)} \#y = (\#Y^*) \sum_{y \in Y \cap T^+(X)} p_y < \frac{\#Y^*}{2}$$

because $x \notin BP(T)$ and because the equilibrium of a Tournament Game is strict (proposition 6.2.8.). Thus Y^* is regular and Copeland dominant for T^*. Then, if S is Copeland-dominant, $Y^* \subset S(T^*)$. But T^* is composed, T being a summary of T^*, thus if S is composition-consistent, $Y \subset S(T)$ and therefore $BP \subset S$.

■

This axiomatization of the Bipartisan set does not add much to our knowledge of the concept because it is merely a re-statement of previous propositions. But it may be worth mentioning because, within the restricted framework of tournament games, it defines a concept for non-cooperative game theory without any reference to the strategic interaction of the players.

6.4. Method of Minimal Gain

In this section, we come back to the problem of the ranking of alternatives by scores, using the idea of the tournament game. Given a tournament T we consider a family of games of which the game $J(T)$ is a particular case.

Definition 6.4.1. : Let $T \in \mathcal{T}(X)$, $x^* \in X$ and $e \in \mathbb{R}^+$. The game $J(T, x^*, e)$ is the mixed strategies two-player zero-sum game with space of pure strategies $X \times X$ and such that if player 1 plays x and player 2 plays y, the payoff for player 1 is :

- 0 if $x = y$
- 1 if $x \rightarrow y$ and $x \neq x^*$
- e if $x \rightarrow y$ and $x = x^*$
- -1 if $x \leftarrow y$ and $y \neq x^*$
- $-e$ if $x \leftarrow y$ and $y = x^*$

In particular, for all $x^* \in X$, $J(T, x^*, 1) = J(T)$. Like $J(T)$, all the games $J(T, x^*, e)$ are symmetric and have symmetric equilibria. Let us denote by $E(J, x^*, e)$ the set of equilibria of $J(T, x^*, e)$. We can write the same characterization as in proposition 6.2.3 (the proof is straightforward) :

<u>Proposition 6.4.2.</u>: Let $p \in \Delta_X$, then $(p, p) \in E(T, x^*, e)$ if and only if, for all $x \in X$:

- if $x \rightarrow x^*$: $p[T^+(x)] \leq p[T^-(x)]$
- if $x \leftarrow x^*$: $p[T^+(x)] \leq p[T^-(x)-\{x^*\}] + ep[\{x^*\}]$
- if $x = x^*$: $e\, p[T^+(x)] \leq p[T^-(x)]$

with, in all cases, equality if $x \in Supp(p)$.

Unlike the case $e = 1$, the games $J(T, x^*, e)$ do not have, in general, unique equilibria:

<u>Proposition 6.4.3.</u>: The set $E(T, x^*, e)$ is non-empty and may contain several elements.

Proof :

The proof of the non-emptiness is standard. Here is an example of multiple equilibria :

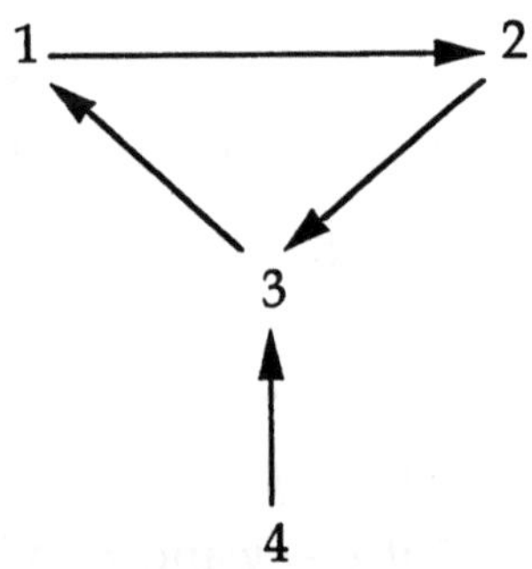

Figure 6.3.

Let T be the tournament of order 4 depicted in figure 6.3. Take $x^* = 4$. Proposition 6.4.2. gives in this case that $(p, p) \in E(T, x^*, e)$ for :

(1) $p_2 + p_4 \leq p_3$

(2) $p_3 + p_4 \leq p_1$

(3) $p_1 \leq p_2 + ep_4$

(4) $ep_3 \leq p_1 + p_2$.

Observe that for $e= 2$, this system has an infinite number of solutions in Δ_X with the four equations at equality, one can verify that these solutions are given by :

$$\begin{cases} p_1 = 1 - 2t \\ p_2 = 4t - 1 \\ p_3 = t \\ p_4 = 1 - 3t \end{cases}$$

with $t \in]\frac{1}{4}, \frac{1}{3}[$. ■

For each value of e, the alternative x^* can appear or not in the supports of the strategies p such that $(p, p) \in E(T, x^*, e)$. Intuitively, an alternative x^* is a good one if it remains played with positive probability when one reduces the payoff when it wins. Hence the idea of the following definition.

<u>Definition 6.4.5.</u>: Let $T \in \mathcal{T}(X)$, $x \in X$. Call *minimal gain* associated to x, and denote by $\eta(x)$ the quantity:

$$\eta(x) = Min\{e \in \mathbb{R} : \forall p \in \Delta_X, (p, p) \in E(T, x^*, e) \Rightarrow p_x > 0\}.$$

Call *minimal gain score* the quantity:

$$\zeta(x) = 1/\eta(x).$$

We are now going to give an explicit formula for the computation of the minimal gain associated to an alternative. This will allow to study the ranking induced by this indicator.

<u>Lemma 6.4.6.</u>: Let $A(x) = \{e \in \mathbb{R} : \forall p \in \Delta_X, (p, p) \in E(T, x, e) \Rightarrow p_x > 0\}$, then $A(x)$ is either empty, either of the form $]\ \eta(x), +\infty\ [$ with $\eta(x) > 0$, either equal to $[0, +\infty[$.

Proof :

Let $e_2 > e_1$, if there exists $p \in \Delta_X$ with $(p, p) \in E(T, x, e_2)$ and $p_x = 0$ proposition 6.4.2 implies that $(p, p) \in E(T, x, e_1)$, thus : $e_1 \in A(x) \Rightarrow e_2 \in A(x)$, so $A(x)$ is equal to $[\eta(x), +\infty[$ or to $]\ \eta(x), +\infty[$. If $\eta(x) > 0$ let $(e^n)_{n>0}$ be a sequence of real numbers strictly smaller than $\eta(x)$, with limit $\eta(x)$ when n tends to infinity, and let $p^n \in \Delta_X$ such that $(p^n, p^n) \in E(T, x, e^n)$ and $p_x^n = 0$. The set Δ_X being compact, one can suppose that p^n tends to a limit p. One can then see that $(p, p) \in E(T, x, \eta(x))$ and that $p_x = 0$, which proves that $\eta(x) \notin A(x)$.

■

It turns out that the minimum gain score $\zeta(x)$ can be computed, with the help of an explicit formula which only involves the equilibrium strategy for the (non-perturbed) game of the sub-tournament T-x.

Proposition 6.4.7.: Let $T \in \mathcal{T}(X)$, $x \in X$. Let $p \in \Delta_{X-\{x\}}$ be the equilibrium strategy for the restriction of T to X-$\{x\}$: $\{(p, p)\} = E(T\text{-}x)$. The minimal gain score of x is :

$$\zeta(x) = p[T^+(x)]/\,p[T^-(x)].$$

Proof :

Let $e \leq \dfrac{p[T^-(x)]}{p[T^+(x)]}$, extend p on X by $p_x = 0$, then proposition 6.4.2. shows that $(p, p) \in E(T, x, e)$, and so, according to the lemma, $e \leq \eta(x)$. This proves that $\eta(x) \geq \dfrac{p[T^-(x)]}{p[T^+(x)]}$. Conversely, let $e > \dfrac{p[T^-(x)]}{p[T^+(x)]}$. Suppose that there exists $(p', p') \in E(T, x, e)$ with $p'_x = 0$, then, reading one more time proposition 6.4.2. one can observe that $(p', p') \in E(T - x)$, hence $p' = p$, which contradicts

$e > \frac{p[T^-(x)]}{p[T^+(x)]}$. As a consequence $\eta(x) \leq \frac{p[T^-(x)]}{p[T^+(x)]}$. The result follows for $\zeta(x) = 1/\eta(x)$.

■

Proposition 6.4.8.: Let $T \in \mathcal{T}(X)$, $x \in X$.

(i) $\zeta(x) > 1 \Leftrightarrow x \in BP(T)$

(ii) $\zeta(x) = 0 \Leftrightarrow BP(T) \subset T^-(x)$

Proof :

(i) $x \in BP(T) \Leftrightarrow 1 \in A(x) \Leftrightarrow \eta(x) < 1 \Leftrightarrow \zeta(x) > 1$

(ii) If $\zeta(x) = 0$ then $p[T^+(x)] = 0$ that is $T^+(x) \cap BP(T - x) = \emptyset$, but $x \notin BP(T)$ thus (theorem 6.3.2. property SSP) $BP(T) = BP(T - x)$, hence $BP(T) \subset T^-(x)$. Conversely if $BP(T) \subset T^-(x)$, $x \notin BP(T)$ thus $BP(T) = BP(T - x)$, thus $p[T^-(x)] = 1$ and $\zeta(x) = 0$.

■

We now make some remarks which clarify the properties of this scoring method.

Remark 6.4.9.: It is possible that $\zeta(x)$ is infinite for all x. One can easily verify that this is the case for the cyclones C_n.

Remark 6.4.10.: It is possible to construct a tournament in which the Copeland winner has a score $\zeta(x)$ equal to zero (construction left to the reader).

Remark 6.4.11.: The score ζ does not behave like the equilibrium probability p. Consider the tournament T_n of order $2n + 2$ defined on $X = \{x_0, 1, ..., 2n + 1\}$ by:

(i) $T_n/\{1, ..., 2n + 1\}$ is isomorphic to C_{2n+1},

(ii) for all $i \in \{1, ..., 2n\}$, $x_0 T_n i$, and

(iii) $(2n + 1)\ T_n x_0$ (see figure 6.4).

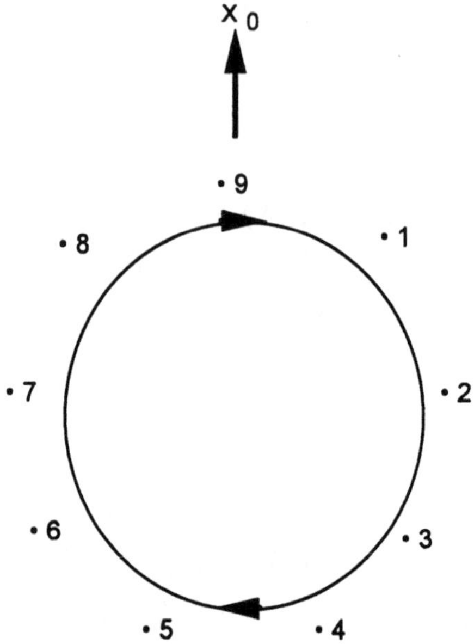

Figure 6.4.: the tournament T_4

In this tournament T_n, x_0 covers the points 1 to n and one has :

$$UC(T_n) = \{x_0, (n + 1), (n + 2), ..., (2n + 1)\}.$$

In $UC(T_n)$, $(n + 1)$ covers $(n + 2)$, $(n + 3)$,... up to $(2n)$, so :

$$UC^2(T_n) = \{x_0, n{+}1 , 2n{+}1)\}.$$

We can deduce that $BP(T_n) = \{x_0, n+1, 2n+1\}$, and we can also deduce the equilibrium strategy, uniform on this set. Let us now compute the scores ζ. The tournament $T_n/X - \{x_0\}$ being cyclical, the equilibrium probability for this tournament is uniform and thus $\zeta(x_0) = 2n$. The tournament $T_n/X - \{2n+1\}$ has x_0 as Condorcet winner, one deduces $\zeta(2n+1) = +\infty$. The tournament $T_n/X - \{n+1\}$ shows a covered alternative : 1 is covered by $2n + 1$. Consequently, $BP(T_n/X - \{n+1\}) = BP(T_n/X - \{1, n+1\})$. But it can be easily checked that $T_n/X - \{1, n + 1\}$ is isomorphic to T_{n-1}, because C_{2n+1} restricted to $\{2, ..., n, n+2, ..., 2n+1\}$ is isomorphic to C_{2n-1}; so the equilibrium strategy for $T_n/X - \{n+1\}$ is uniform on $\{x_0, 2n+1, n\}$, so $\zeta(n+1) = 2$. The reader will verify that for any other point i the score is $\zeta(i) = 1/2$.

Remark 6.4.12.: Let $x \in X$ and $y \in X$ be such that yTx. Let us denote by $\zeta(x)$ and $\zeta'(x)$ the minimal gain scores of x in the tournaments T and $T_{<x,\, y>}$. The restrictions to $X - \{x\}$ of T and $T_{<x,\, y>}$ are identical, and so one may write: $\zeta(x) = \dfrac{p[T^+(x)]}{p[T^-(x)]}$ and

$\zeta'(x) = \dfrac{p[T^+(x)] + p(\{y\})}{p[T^-(x)] - p(\{y\})}$, hence $\zeta'(x) \geq \zeta(x)$. This monotonicity property makes the interpretation of $\zeta(x)$ as a "strength" of x in the tournament is attractive (compare with example 6.2.9.).

Example 6.4.13.: To illustrate the method, the minimal gain scores will now be computed for the tournament which was analysed in chapter 4 (this tournament is depicted in figure 4.1). for this tournament T, the subset $\{2,3\}$ of $X = \{1, 2, 3, 4, 5, 6, 7\}$ is a component and it is easy to check that there are no other non-trivial component. Let T' be the summary of T. By looking at T' one finds immediately the covering relation for T:

- 2 covers 3. 5 and 6
- 4 covers 7,

and the Uncovered set is $UC(T) = \{1, 2, 4\}$. From the properties of the Bipartisan set, it follows that the Bipartisan set, the Minimal Covering set as well as the Iterated Uncovered set also reduce to $\{1, 2, 4\}$, and that the vector of Nash equilibrium probabilities for T is $(1/3, 1/3, 0, 1/3, 0, 0, 0)$.

Consider now the restricted tournaments $T\text{-}x$ for $x \in X$.

- $T\text{-}1$: In this tournament, 2 is a Condorcet winner, therefore then Nash equilibrium places a unit mass on 2 and, since *1* beats 2 the minimal gain score of *1* is : $\zeta(1) = +\infty$.
- $T\text{-}2$: Since $\{2, 3\}$ is a component of T, the tournament $T\text{-}2$ is isomorphic to the summary T'. Therefore the Nash equilibrium for $T\text{-}2$ places the probability 1/3 on *1*, 3 and 4. Because 2 beats 3 and 4 and is beaten by *1*, the minimal gain score of 2 is : $\zeta(2) = 2$.
- $T\text{-}3$: By the same reasoning, $\zeta(3) = 1/2$.

- *T-4* : In this tournament, *3, 5* and *6* are covered and the Bipartisan set is *{1, 2, 7}* therefore $\zeta(4) = 2$.
- *T-5* : *6* and *7* are covered in *T-5* thus $\zeta(5) = 0$.
- *T-6* : $\zeta(6) = ½$.
- *T-7* : $\zeta(7) = ½$.

The vector of the minimal gain scores for *T* is thus :

$$\zeta = (+\infty, 2, ½, 2, 0, ½, ½).$$

6.5. Interpretation of Tournament Games

The class of tournament games is a class of normal-form games on which many concepts of non-cooperative game theory have particular clear expression (cf. table 3 in the annex). For instance, dealing with tournament games, there is no need to distinguish large and strict domination. Existence and unicity hold for Shapley's Weak Saddle as well as for the mixed Nash equilibrium.

These games can be seen as "qualitative decisive duels": they are two-player symmetric zero-sum games in which the payoffs are qualitative, except on the diagonal, $g(x, y)$ is either "good" (+1) or "bad" (-1). They generalize the game "Paper, Rock, Scissors" played by children, in which Paper smothers Rock, Rock smashes Scissors, and Scissors cut Paper". This game is the tournament game of a 3-cycle.

One interpretation of a tournament game is a story of medieval warriors. Two knights are fighting in a ... tournament and are allowed weapons: sword, clubs, mace, flail, ... The two knights being equally brave and skilled, the winner only depends upon the chosen weapons. The knights' problem is to choose a weapon. Another interpretation is a more modern contest: two political parties make proposals and voters vote on the basis of these proposals. Suppose that the individual's preferences are strict and that there is an odd number of voters, then, if the winner of the election is the party which gets the majority of the votes (whatever the size of this majority is), the two-player game is the tournament game $G(T)$ where T is the majority tournament of the voter's preference profile (*cf.* 2.1). We shall come back to this interpretation in chapter 10.

The class of tournament games is a nice class of games for picking examples and counter-examples of games. For instance, the remark that the Iterated Uncovered set is not monotonous (theorem 5.2.4. (i)) is a strange property of the process of iterative elimination of dominated strategies. It is also sometimes possible to obtain theorems on some classes of games including the tournament games by rewriting the argument originally designed for tournaments. Two examples of this feedback from Graph Theory to Game Theory are the following.

The proof of the existence and uniqueness of a minimal covering set, originally given by Dutta (1988), is extended by Duggan and LeBreton (1996) who obtain the following theorem.

<u>Theorem 6.5.1.</u>: Any finite, symmetric, two-player, zero-sum game (X, g) such that

$$\forall (x, y) \in X^2, x \neq y \Rightarrow g(x, y) \neq 0$$

has a unique Weak Saddle.

The proof of the existence and uniqueness of the mixed Nash equilibria of a tournament game, originally given by Laffond, Laslier and LeBreton (1993a), is extended by the same authors (1996) who obtain:

<u>Theorem 6.5.2.</u>: Any finite, symmetric, two-player, zero-sum game (X, g) such that

$$\forall (x, y) \in X^2, x \neq y \Rightarrow g(x, y) \text{ is an odd integer}$$

has a unique mixed Nash equilibrium.

More on these interpretations and generalizations will be mentionned in chapter 10.

7 - The Contestation Process

Let S be a tournament solution and $T \in \mathcal{T}(X)$. Consider the binary relation on X denoted by $D(S,T)$ and defined by :

$$\forall(x,y) \in X^2, xD(S,T)y \Leftrightarrow x \in S(T \mid T^-(y)).$$

If $xD(S,T)y$, we shall say that x is a *contestation* of y for T according to S. Intuitive justification for this name is the following: Imagine that outcome y is under consideration for being a "good" one. The arguments against y being a good outcome are to be found in the outcomes which beat y, that is the set $T^-(y)$, so we consider the tournament T restricted to the predecessors of y. Among these outcomes, the best ones are chosen according to S, that is to say, the outcomes which are to be opposed to y are the outcomes $x \in S(T \mid T^-(y))$. As a particular case, if there is no such x for a given y, then the set of predecessors of y must be empty, thus y is a Condorcet winner and is unique. The relation of contestation (which is a sub-relation of T) has been introduced by Schwartz (1990) who uses it in order to define the "Tournament Equilibrium set" presented in the second section of this chapter. It is also possible to use the contestation relation in order to define the solution proposed by Banks (1985) and studied in particular by Bordes (1988) and Miller, Grofman and Feld (1990).

7.1. Banks' Solution

We shall use the following notations, which generalize those adopted for tournaments : if U is a binary relation on X and $x \in X$, the sets of successors and predecessors of x are respectively denoted by $U^+(x) = \{y \in X : x \, U \, y\}$ and $U^-(x) = \{y \in X : y \, Ux\}$. For a subset Y of X, we shall also denote

$$U^+(Y) = \bigcup_{x \in Y} U^+(x), \text{ and } U^-(Y) = \bigcup_{x \in Y} U^-(x),$$

and in particular, we shall often consider the sets $U^+(X)$ and $U^-(X)$. For instance if S is a tournament solution $D(S,T)^-(X)$ is the set of points of X which are contestation of some point of X according to S.

Proposition 7.1.1.: There exists a unique tournament solution, B, such that

$$\forall T \in \mathcal{T}(X), \, o(T) \geq 2 \Rightarrow B(T) = D(B,T)^-(X).$$

Proof :

We prove by induction on the order of T that B is uniquely defined. For tournaments of order 1, $B(T) = X$ by definition of a tournament solution; suppose that B has been defined for all the tournaments of order smaller or equal to n, and let $T \in \mathcal{T}(X)$ be a tournament of order $n{+}1$, then for any $x \in X$, $T^-(x)$ has a cardinal smaller or equal to n, thus $B(T^-(x))$ is well defined by $B(T) = D(S,T)^-(X)$. It is straightforward to verify that B is a tournament solution. ■

This solution, which we shall call the *Banks' solution,* has been introduced differently, while working on the resolution of tournaments by the mean of binary trees; we will come to this point in the next chapter. Usually, the solution B is defined by the following characterization, due to Banks (1985).

<u>Proposition 7.1.2.:</u> Let $x \in X$ and $T \in \mathcal{T}(X)$, then $x \in B(T)$ if and only if there exists $Y \subset X$ such that $x \in Y$, $T|Y$ is an ordering whose x is the winner, and no point of X beats all the points of Y.

Proof :

By induction on the order n of the tournament. For $n=1$ it is obvious. For $n>1$, if x is the winner of a transitive sub-tournament $T|Y$, let $y \in Y$ be the loser of $T|Y$, then $T|Y - \{y\}$ is maximal transitive in $T^-(y)$ thus by the induction hypothesis $x \in B(T^-(y))$, so $x \in B(T)$. Conversely, if $x \in B(T)$ there exists $y \in X$ such that x contests y according to B, that means $x \in B(T^-(y))$ so by the induction hypothesis, x is the winner of a maximal transitive sub-tournament $T^-(y)|Y$ of $T^-(y)$, one then can see that $T|(Y \cup \{y\})$ is transitive, maximal in X and has x as a winner. ■

We use the expression "maximal transitive chain" for the sets Y such that $T|Y$ is an ordering and such that no point of X dominates all the points of Y. Let us now come to the properties of the Banks' solution.

<u>Theorem 7.1.3.:</u> The Banks' solution, B

(i) is monotonous

(ii) is not independent of losers

(iii) does not verify SSP

(iv) is not idempotent

(v) satisfies the Aïzerman property

(vi) is composition-consistent and thus is included in the Top-Cycle

(vii) is not regular.

Proof :

(i) Clearly if x is the winner of a maximal transitive chain of T it remains so if one reinforces it.

(ii) Consider the tournament T_1 described in figure 5.1. It is easily seen that $B(T_1) = \{1,2,3\}$. Modifying the sub-tournament $\{1',2',3'\}$ into a transitive chain, one gets the tournament T'_1 described in figure 7.1 :

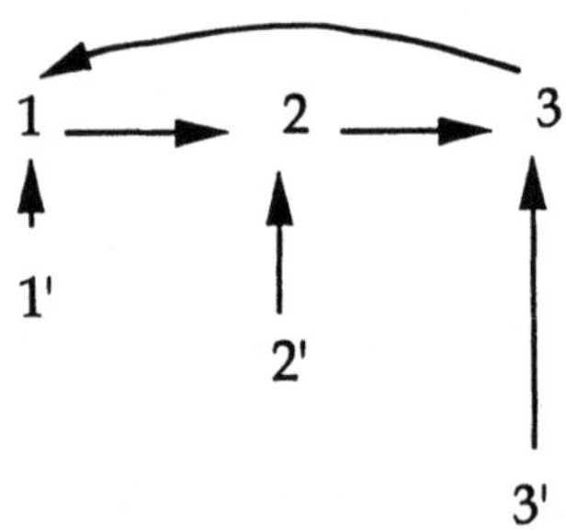

Figure 7.1.

For this tournament, $B(T'_1) = \{1,2,3,1'\}$.

(iii) and (iv) For this same tournament, $B^2(T'_1) = \{1,2,3\}$.

(v) Let $T \in \mathcal{T}(X)$, and let $Y \subset X$, with $B(X) \subset Y$ (We write $B(X)$ for $B(T)$ and $B(Y)$ for $B(T \mid Y)$). Let $y \in B(Y)$ and let $Z \subset Y$ be a maximal transitive chain in Y, with y as a winner. If Z is not maximal in X there exists $x \in X$ such that $x \to Z$, then there exists $x' \in B(X)$ such that $x' \to Z$ and because $B(X) \subset Y$ this contradicts the maximality of Z in Y. Therefore Z is maximal in X and $y \in B(X)$. Hence $B(Y) \subset B(X)$.

(vi) Let $T \in \mathcal{T}(X)$ be a composed tournament, $\widetilde{X} = \{X_1,\ldots,X_n\}$ be a decomposition of T and $T_i = T/X_i$. Let $x \in B(T)$, then there is a maximal transitive chain Y with x as the winner. Let us denote $Y_i = Y \cap X_i$, then it is easily seen that the set of X_i such that $Y_i \neq \varnothing$ forms a transitive chain of the summary tournament $\widetilde{T}$, maximal in $\widetilde{X}$, and the winner of which is $X_{(x)}$ the component to which x belongs. The set $Y_{(x)}$ also forms a transitive chain of $T|X(x)$, maximal in $X_{(x)}$, and having x as winner. This proves the inclusion

$$B(T) \subset \cup\{ B(T_i) : X_i \in B(\widetilde{T})\}.$$

Conversely, let $\widetilde{Y}$ be a maximal transitive chain of $\widetilde{T}$, X_i be the winner of $\widetilde{Y}$, and let Y_i be a maximal transitive chain of T_i and x_i be the winner of Y_i. For any $X_j \in \widetilde{Y}$, different from X_i, let x_j be a point of X_j, and consider the set $Y = Y_i \cup \{x_j \in \widetilde{Y}, j \neq i\}$, one can verify that Y is a transitive chain for T, maximal in X and having x_i as its winner, thus $x_i \in B(T)$, this proves the converse inclusion

$$\cup\{ B(T_i) : X_i \in B(\widetilde{T})\} \subset B(T).$$

(vi) The counter-example is given in 7.1.4.

■

To give the relations between B and the other solutions, we will mainly use one example and one lemma. The rather complicated example is taken from Laffond and Laslier (1991).

<u>Example 7.1.4.:</u>

Consider the tournament Q_7 of the quadratic residues modulo 7 (proposition 1.2.2). In this tournament, the relations are : $1 \to 2,3,5$; $2 \to 3,4,6$; etc.. It is not difficult to establish the list of all the maximal transitive chains of Q_7, there are 21 such chains, each of them has three points. They can be deduced from the three chains $\{1,2,3\}$, $\{1,3,5\}$, $\{1,5,2\}$ by the tournament automorphism $\varphi(i) = i+1$. We shall denote:

$\alpha_{1,1} = 123,$	$\alpha_{2,1} = 135,$	$\alpha_{3,1} = 152$
$\alpha_{1,2} = 234,$	$\alpha_{2,2} = 246,$	$\alpha_{3,2} = 263$
.	.	.
.	.	.
.	.	.
$\alpha_{1,7} = 712,$	$\alpha_{2,7} = 724,$	$\alpha_{3,7} = 741$

those 21 chains. Also we shall denote $A_i = \{\alpha_{i,j} : j = 1,...7\}$ and $A = A_1 \cup A_2 \cup A_3$. Remark that each point of Q_7 belongs to exactly 9 chains $\alpha_{i,j}$.

For any two integers n and m, we are going to build a tournament $T_{n,m}$ of order $21+7(2n+1)+(2m+1)$ on $\boldsymbol{A} \cup \boldsymbol{B}_n \cup \boldsymbol{C}_m$, where $\boldsymbol{A}$ is the set that has just been described, $\boldsymbol{B}_n = \{\beta_{k,l} : k \in \{1,...,7\}, l \in \{1,...,2n+1\}\}$ and $\boldsymbol{C}_m = \{c_i : i \in \{1,...,2m+1\}\}$.

- Inside $\boldsymbol{A}$: $T|A_i$ is cyclical of order 7, $A_1 \to A_2$, $A_2 \to A_3$ and $A_3 \to A_1$.
- Inside $\boldsymbol{B}_n$: Let $B_{i,n} = \{\beta_{ij} : j \in \{1,...,2n+1\}$, $T|_{B_{i,n}}$ is cyclical of order $2n+1$ and $T|_{B_n} = \Pi(Q_7; C_{2n+1},...,C_{2n+1}\}$.
- Inside $\boldsymbol{C}_m$: Cyclical of order $2m+1$.
- Between $\boldsymbol{A}$ and $\boldsymbol{B}_n$: $\alpha_{ij} \to \beta_{k,l}$ iff $k \in \alpha_{ij}$.
- Between $\boldsymbol{A}$ and $\boldsymbol{C}_m$: $\boldsymbol{A} \to \boldsymbol{C}_m$.
- Between $\boldsymbol{B}_n$ and $\boldsymbol{C}_m$: $\boldsymbol{C}_m \to \boldsymbol{B}_n$.

The reader will observe that the tournament $T_{n,m}$ so defined is composed and admits as a summary the tournament $T_{o,o}$ of order $21+7+1=29$. Let us find the Banks set of the tournament $T_{o,o}$.

Let α_{ij} be a point of $\boldsymbol{A}$, α_{ij} dominates in $\boldsymbol{B}_o$ a chain (b_1,b_2,b_3) and the chain $(\alpha_{ij}, b_1, b_2, b_3)$ is maximal in $T_{o,o}$, thus $\boldsymbol{A} \subset B(T_{o,o})$. One can also see that $\beta_{1,1} \to \beta_{2,1} \to \beta_{3,1} \to 567$ constitutes a transitive chain, which is maximal because $(\beta_{1,1}, \beta_{2,1}, \beta_{3,1})$ is maximal in B, the only point 123 of A beats this chain and $567 \to 123$. Thus $\beta_{1,1}$ belongs to the Banks set. But the isomorphism $\varphi(i)=i+1$ of Q_7 induces an isomorphism on $T_{n,m}$ and so $\boldsymbol{B}_o \subset B(T_{o,o})$. The transitive chains beginning in $\boldsymbol{C}_o$ have no point of $\boldsymbol{A}$ and are followed in $\boldsymbol{B}_o$ by transitive chains of $\boldsymbol{B}_o$, thus, by construction, these chains are beaten by a point of $\boldsymbol{A}$; this proves that $\boldsymbol{C}_o \cap B(T_{o,o}) = \varnothing$. Hence we have $B(T_{o,o}) = \boldsymbol{A} \cup \boldsymbol{B}_o$, and by composition-consistency, for all n and m, $B(T_{n,m}) = \boldsymbol{A} \cup \boldsymbol{B}_n$.

Counting the Copeland scores in $T_{n,m}$ gives :

1) The score of each point of A is: $s_A = 10+3(2n+1)+2m+1$
2) The score of each point of B_n is: $s_B = 12+n+3(2n+1)$
3) The score of each point of C_m is: $s_C = 7(2n+1)+m$.

Take now $n=1$ and $m=1$, then $s_A = s_B = s_C = 22$, thus $T_{1,1}$ is regular, which proves the result announced in 7.1.4 : the Banks solution is not regular. The same example gives the following result.

Proposition 7.1.5: (i) $C \varnothing B$
(ii) $SL \varnothing B$.

Proof:

Take $n=3$ and $m=2$ in example 7.1.4. Then $s_A=36$, $s_B=36$ and $s_C=51$, thus $C(T_{3,2})$ = C which proves (i); moreover $o(T_{3,2}) = 75$ and we know (proposition 3.4.3) that a Slater's winner beats at least half of the points, thus $SL(T_{3,2})$ = C, which proves (ii). ■

It is possible to find smaller counter-examples in order to prove proposition 7.1.5. Charon, Hudry and Woigard (1996) mention a tournament of order 13 with a Slater winner outside the Banks set, and a tournament of order 16 such that the two sets $SL(T)$ and $B(T)$ have empty intersection. Although the orders of these tournaments are smaller than example 7.1.4., computation of their Slater sets is more difficult. For small tournaments, Slater winners are always in the Banks set but the order of the smallest counter-examples is not known.

In order to compare the Banks solution with the solutions introduced in the chapter 5 (covering), the following lemma is useful.

Lemma 7.1.6.: If Y is a covering set of T then $B(T \mid Y) \subset B(T)$.

Proof:

Let $y \in B(T \mid Y)$ and let A be a maximal transitive chain in Y with winner y. If $y \notin B(T)$ there exists $z \in T$ such that $z \to A$. Because A is maximal in Y, $z \notin Y$; then (cf. definition 5.3.1.) $z \notin UC(T \mid Y \cup \{z\})$ implies that there exists $x \in Y$ covering z and thus beating all the points of A, a contradiction. ■

Proposition 7.1.7.: (i) $B(MC) \subset B(UC^{\infty}) \subset B(UC^{k}) \subset B$

(ii) $B \cap MC \neq \varnothing$.

Proof:

Point (i) is immediately deduced from the preceding lemma and (ii) is deduced from (i). ■

Proposition 7.1.8: (i) $B \subset TC(UC) \subset UC$

(ii) $B \not\subset UC^2$

(iii) $B^k \subset UC^k$ and $B^{\infty} \subset UC^{\infty}$.

Proof:

(i) See Banks (1985).

(ii) The counter-example is the one in figure 5.5. One verifies that for this tournament T_1, $B(T_1) = UC(T_1) = \{x_1, x_2, x_3, x_4\}$ and $UC^2(T_1) = \{x_1, x_2, x_3\}$

(iii) This point is deduced from propositions 2.3.8. and 7.1.3.(v) (Aïzerman property) and from the inclusion $B \subset UC$.

■

The solution B intersects the successive refinements of the Uncovered Set, down to the Minimal Covering Set, but is not included in UC^2, thus it remains, for completeness of the analysis, to know if there is an inclusion relation between BP and B and, if not, if BP and B intersect. We only have the answer to the first of these questions.

<u>Proposition 7.1.9:</u> $\exists T \in \mathcal{T}_{29} : B(T) \subset BP(T)$ and $B(T) \neq BP(T)$.

Proof :

Take again example 7.1.4 : $T_{0,0}$ is of order 29, and summary of $T_{1,1}$, but $T_{1,1}$ is regular, hence $BP(T_{0,0}) = \boldsymbol{A} \cup \boldsymbol{B_0} \cup \boldsymbol{C_0}$; but we know that $B(T_{0,0}) = \boldsymbol{A} \cup \boldsymbol{B_0}$. ■

<u>Proposition 7.1.10:</u> $\exists T \in \mathcal{T}_6 : BP(T) \subset B(T)$ and $B(T) \neq BP(T)$.

Proof :

The counter-example is the one of proposition 6.3.3. (figure 6.2.). One has $B(T) = \{1,2,3,4,5,6\}$, maximal transitive chains are for this tournament $(1,2,4)$, $(2,6,4)$, $(3,6,1)$, $(4,5,3)$, $(5,3,2)$ and $(6,1,4,5)$.

■

It has been observed that the Banks set is a composition-consistent refinement of the solution $TC(UC)$. Therefore, it is possible that the Banks set helps locating the solution SL^*, defined as the composition-consistent hull of the Slater solution. We know that, like B, SL^* is a refinement of $TC(UC)$ but not of UC^2. Since Slater winners may not be in the Banks set it is clear that $B \neq SL^*$, but it is true that $B \subset SL^*$, as stated in the following result. This shows that any point in the Banks set can become a Slater winner for some cloning of the outcomes (see propositions 2.5.4. and 5.1.9.). We first need a lemma concerning SL^*.

<u>Lemma 7.1.11.:</u> Let $T \in \mathcal{T}(X)$ and $y \in X$. For any positive integer n, consider a set $Y_n = \{y_1, ..., y_n\}$ of n elements such that $X \cap Y_n = \emptyset$. Denote $X_n = X - \{y\} \cup Y_n$, and let T_n be the tournament on X_n such that :

1) $T = T_1$ is a summary of T_n, with y in place of Y_n: $T_n/(X_n - Y_n) = T/(X - \{y\})$ and for all $x \neq y$, $xT_ny_i \Leftrightarrow xTy$, $i = 1, \ldots, n$.

2) T_n/Y_n is linear, with y_1 on top.

Then, for n sufficiently large, any Slater order U_n for T_n is such that $T^-(y)$, Y_n and $T^+(y)$ are intervals for U_n, with $T^-(y)\ U_n\ Y_n\ U_n\ T^+(y)$.

Proof :

Let U_n be such that $T^-(y)$, Y_n and $T^+(y)$ are intervals for U_n, and the restriction of U_n to these intervals are Slater orders of the associated sub-tournaments. Then one has: $\Delta\ (U_n, T_n) = \Delta^+ + \Delta^- + \Delta^{+-}$, with:

Δ^+ is the Slater distance for $T/T^+\ (y)$,

Δ^- is the Slater distance for $T/T^-\ (y)$,

$\Delta^{+-} = \#\{(x, x') \in T^-(y) \times T^+(y): x'Tx\}$,

These three numbers being independent of n. It is easy to verify that for n large enough, U_n is a Slater order for T_n. ■

Proposition 7.1.12.: $B \subset SL^*$.

Proof :

By induction of the order of the tournament. Let $T \in \mathcal{T}(X)$, if T has order 1, it is trivial. If $o(T) \geq 2$, let $x \in B(T)$, from proposition 7.1.1. there exists $y \in X$ such that $x \in B(T^-(y))$. By the induction hypothesis, $x \in SL^*(T^-(y))$. Thus there exists a composed tournament $\tilde{T}$, with $T/T^-(y)$ as a summary, such that the component x is on top of a Slater order for $\tilde{T}$. From lemma 7.1.11., by cloning y, it is possible to construct a tournament T' such that $\tilde{T} = T'/T'^-(y)$ and the component x is on top of a Slater order for T'. Then T is a summary of T' and therefore composition-consistency implies that $x \in SL^*(T)$.

■

The Banks set has some non-trivial properties related to the notion of covering, but is not a refinement of the Minimal Covering set. Since the Minimal covering set can be axiomatized with the help of the weak expansion properties γ^* or γ^{**} (cf. 5. 3), one may wonder whether the Banks set satisfies these properties. The answer is negative.

Proposition 7.1.13.: The Banks set does not satisfy properties γ^* or γ^{**}.

Proof :

Let $Y = \{1, 2, 3, 4, 5, 6, 7, 8, 9\}$ and $X = Y \cup \{0\}$. Let T the tournament on X be defined by:

- T/Y is a cyclone of order 9, with $1 \rightarrow 2345$, $2 \rightarrow 3456$, *etc...*
- $0 \rightarrow 147$ and $235689 \rightarrow 0$.

It is easy to verify that $B(T) = Y$ although $UC(T) = X$. Therefore, γ^{**} (and thus γ^*) is not satisfied by B. ■

7.2. The Tournament Equilibrium Set

This solution, which is a refinement of the solution B has been introduced by Schwartz (1990) and is called the *Tournament Equilibrium Set* (TEQ). In order to define it, we use the concept of Top-Set which was defined in the first chapter (definition 1.5.10.). Recall that $\mathcal{R}(X)$ denotes the set of binary relations on X. For $R \in \mathcal{R}(X)$ and $Y \subset X$, Y is retentive for R if there exist no $y \in Y$ and $x \in X - Y$ such that xRy, and the Top-Set of R, $TS(R)$, is the union of the minimal retentive subsets of R. If $R = D(S,T)$ is the relation of contestation associated to a solution S on T, $TS(R)$ is the minimal set of points into which one ends by iterating the process of contestation. Hence the definition of Schwartz's set :

Proposition 7.2.1.: There exists a unique tournament solution, TEQ, such that

$$\forall T \in \mathcal{T}(X),\ TEQ(T) = TS\,[D(TEQ,T)].$$

Proof :

Straightforward, by induction on the order of T. ■

Following the original presentation by Thomas Schwartz, one can define TEQ in an axiomatic way by the three following axioms; for a tournament solution S, call S-retentive any subset retentive for $D(S,T)$, where $T \in \mathcal{T}(X)$.

Retentiveness axiom : $S(T)$ is S-retentive.

Predictability strength axiom : there is no strict subset Y of $S(T)$ such that Y is S-retentive and $S(T)-Y$ is not.

Inclusiveness axiom : if $Y \subset X$ is S-retentive then $Y \cap S(T) \neq \varnothing$.

It is clear that this presentation is equivalent to proposition 7.2.1., because in fact the definition of the Top-Set (1.5.10.) summarizes the three axioms of Schwartz. Comparing the definitions of B and TEQ (7.1.1. et 7.2.1.) one can see that TEQ is a refinement of B (this fact is proved by Schwartz (1990) using the characterization of the Banks set by maximal transitive chains).

Proposition 7.2.2.: $TEQ \subset B$.

Proof :

By induction on the order of T. If $o(T) = 1$ it is obvious. Moreover, if $x \in TEQ(T)$, x belongs to the Top-Set of $D(TEQ,T)$, in particular there exists $y \in X$ such that $x \in TEQ(T^-(y))$. But $T^-(y)$ has an order smaller than T, thus $x \in B(T^-(y))$, thus $x \in B(T)$.

■

The following propositions give some precision concerning the localization of TEQ but does not completely solve the question. For instance it is not known whether TEQ is included or not in MC, neither is it known if the intersection $TEQ \cap BP$ may be empty.

<u>Proposition 7.2.3.:</u> $TEQ \circ UC^\infty = TEQ$.

Proof :

Let $T \in \mathcal{T}(X)$, according to 5.2.3 it is sufficient to prove that if $x \in UC(T)$, $TEQ(T) = TEQ(T|_{X-\{x\}})$, which will be done by induction on the order of T. If the order of T is 1 it is trivial. Let $T \in \mathcal{T}(X)$ be a tournament of order $n>1$ and $x \in UC(T)$. Denote $X' = X-\{x\}$ and $T' = T - x = T|_{X'}$. Let $y \in X'$, if $x \notin T^-(y)$, then clearly $TEQ(T^-(y)) = TEQ(T'^-(y))$, if $x \in T^-(y)$, the point which covers x in X *a fortiori* covers x in $T^-(y)$, hence, by the induction hypothesis, $TEQ(T^-(y)) = TEQ(T'^-(y))$. The relations $D(TEQ,T)$ and $D(TEQ,T')$ are thus identical on $X-\{x\}$ and moreover there exists no point $y \in X$ such that $xD(TEQ,T)y$. Since $T^-(x) \neq \emptyset$ there exists $y \in X$ such that $y\,D(TEQ,T)x$, and we get $TEQ(T) = TEQ(T')$. ■

<u>Proposition 7.2.4.:</u>

(i) $\exists T \in \mathcal{T}_8 : TEQ(T) \subset MC(T)$ and $TEQ(T) \neq MC(T)$

(ii) $\exists T \in \mathcal{T}_{29} : TEQ(T) \subset BP(T)$ and $TEQ(T) \neq BP(T)$

(iii) $\exists T \in \mathcal{T}_6 : BP(T) \subset TEQ(T)$ and $TEQ(T) \neq BP(T)$

(iv) $\exists T \in \mathcal{T}_9 : TEQ(T) \subset B(T)$ and $TEQ(T) \neq B(T)$.

Proof :

(i) See Dutta (1990).

(ii) Is deduced from 7.1.9. and 7.2.2.

(iii) The counter-example is the one of proposition 6.3.3, which has been also used for proposition 7.1.10. (see figure 6.2.). The relation $D(TEQ,T)$ is represented on figure 7.2., of course this relation is not a tournament, and all the arrows have been depicted. One can verify that $TEQ(T) = \{1, 2, 3, 4, 5, 6\}$, and it has been observed that $BP(T) = \{1, 2, 3, 4, 6\}$.

(iv) See Schwartz (1990).

■

The next proposition states that the Tournament Equilibrium set (and the Banks' set) of a tournament without a Condorcet winner has at least three elements. The same result was proven for the Bipartisan set (and its supersets) in proposition 6.3.7.

Proposition 7.2.5.: The Tournament Equilibrium set of a tournament T is a singleton $\{x\}$ if and only if x is Condorcet winner of T, and if T has no Condorcet winner then $TEQ(T)$ contains at least three points. The same statement is true for the Banks' set.

Proof :

If x is not a Condordet winner then $T^{-}(x) \neq \varnothing$ hence there exists y such that $yD(TEQ, T)x$ and $TEQ(T) \neq \{x\}$. It follows that $TEQ(T) = \{x\}$ if and only if x is Condorcet winner. Suppose that $TEQ(T) = \{x, y\}$ with xTy then y is not a contestation of x and since $TEQ(T)$ is retentive, x has in fact no contestation thus x is Condorcet winner. This prove that $TEQ(T)$ cannot have two elements.

The proposition for the Banks' set follows from the inclusion $TEQ \subset B$. ■

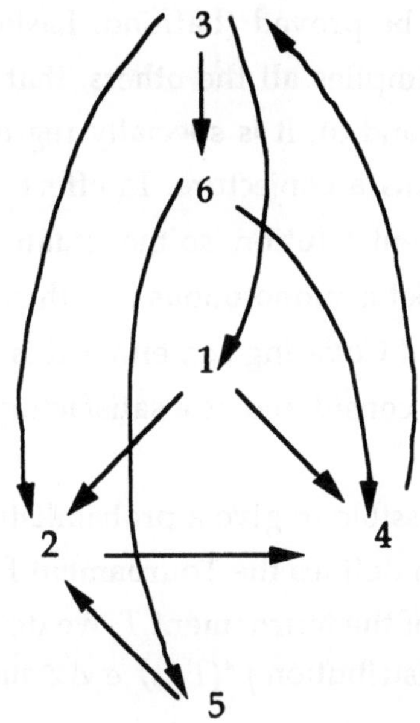

Figure 7.2. :
Relation of contestation for the tournament of figure 6.2.

The solution *TEQ* looks promising, unfortunately few things are known concerning its properties. The following conjectures can be made.

Conjectures 7.2.6.: 1) *TEQ(T)* is absorbing for *D(TEQ,T)*, which means that for any $x \in X$ there exists $y \in TEQ(T)$ such that $yD(TEQ,T)x$.

2) There is only one minimal retentive subset for *D(TEQ,T)*.

3) *TEQ* is monotonous.

4) *TEQ* satisfies the Strong Superset Property.

5) *TEQ* is independent of the losers.

6) $TEQ \subset MC$.

These conjectures are not independent the ones from the others. It can be proved (Laffond, Laslier and LeBreton 1993b) that the first one implies all the others, that 2), 3) and 4) are equivalent and imply 5) and 6). It is specially regrettable that the monotonicity of *TEQ* remains a conjecture. In effect, this property seems natural for a tournament solution, so the situation is : either the Tournament Equilibrium Set is monotonous and then it is a very good refinement of the Minimal Covering set, either it is not monotonous and then it can hardly be considered as a satisfactory tournament solution.

It is possible to give a probabilistic version of the contestation process which defines the Tournament Equilibrium set. By induction on the order of the tournament T, we define on the set X of vertices a probability distribution $p^*(T, .) \in \Delta_X$ such that $p^*(T,x)>0$ if and only if $x \in TEQ(T)$.

If $n=1$, or if T has a Condorcet winner, $p^*(T, .)$ is trivial. If $n>1$ and T has no Condorcet winner suppose that we know $p^*(T', .)$ for all the tournaments T' of order less than n and let T be a tournament of order n without a Condorcet winner. Let $x \in X$, $T^-(x)$ has an order smaller than n ; for any $y \in X$, let $P(T; x, y) = 0$ if $x=y$ or if xTy, and $P(T; x, y)=p^*(T^-(x), y)$ if yTx. Then $P(T; x, y) > 0$ if and only if $yD(TEQ,T)x$. Now we have to find the Top-Set of this relation, a straightforward solution is the following. Consider a Markov chain on X with transition probabilities $P(T; . , .)$; let $p^{(t)}$, $t \geq 0$, be a sequence of probability distributions on X such that for all $y \in X$,

$$p^{(t+1)}(y) = \Sigma_{x \in X} P(T;x,y) p^{(t)}(x),$$

and take initially $p^{(0)}$ uniform on X. Then this sequence has a limit, the support of which is the union of the *TEQ*-retentive subsets of T, and we define $p^*(T, .)$ as this limit.

Observe that the values of $p^*(T, x)$ depend on the values of the initial probabilities $p^{(0)}$, except if there is only one minimum retentive subset. But this is precisely one of the related conjectures about the Tournament Equilibrium Set (the second one). This means that if the conjectures mentioned above are true, then there is a natural scoring method associated with the Tournament Equilibrium set.

Example 7.2.7.: These probabilities - call them the *TEQ-scores* - exist for the tournament of order 7 used as an example in chapter 4 (cf figure 4.1 and example 6.4.13) because the statements in the conjectures 7.2.6 are true for the tournaments of small orders. It is not difficult to compute them. Since the Uncovered set of this tournament is made of three points *{1, 2, 4}*, the Banks' set and the Tournament equilibrium set also reduce to that same set. It is straightforward to deduce that, in such a case, the *TEQ*-scores are 1/3 for the alternatives *1*, *2* and *4* and 0 for the other alternatives. Results concerning this tournament are gathered in the annex.

8 - Tournament Algebras and Binary Trees

One familiar way of determining winners in sport competitions is to schedule a series of games between the participants according to some fixed schedule or *agenda,* in such a manner that fewer and fewer participants remain along the competition, up to the last game, called the final, the winner of which is said to have won the competition. These schedules are not *round-robin tournaments* because it is not the case that each possible pairwise game is played once and only once, but if one assumes that there exists an underlying tournament providing without ambiguity the result of all possible games then studying the properties of various agendas becomes an interesting exercise in Tournament Theory.

The most common example of an agenda is the direct balanced eliminination schedule, sometimes called "knockout tournament" in which 2^n participants play successive rounds called n^{th} final, ..., quarterfinal, semifinal, final, half of the contestants being eliminated in each round. Here, more general schedules will be considered, including the possibility for some players to come back after a loss.

The definition by means of binary tree of a schedule of binary contests, or agenda, is given in section 2. In order to schedule the competition for a given set X of n players, one needs to have an agenda suitable for n players (a *board* of order n) and then to specify a *drawing*, that is to assign to each player his or her initial place in the schedule. This material being set, each tournament on X defines the results of all games and a winner for the competition. Since that winner depends both on the tournament and on the drawing, this procedure does not define a tournament solution but a (non-neutral) choice function. In order to study the properties of the agendas independently of the drawings, one has to consider the set of all the possible winners with all the possible drawings. Doing so, an agenda defines a tournament solution.

The basic element when solving tournaments by binary trees is the tournament *operation* which associates to any pair (x,y) of players the winner of the game played between x and y. Therefore it is not a suprising fact that the terminology of operation algebra is more suitable for expounding on this matter than the terminolgy of relation algebra or graph theory (although they are logically equivalent). Indeed, it turns out that the tournament solutions computable by binary trees are precisely those defined by algebraic functions of n variables (the analog of polynomials for a non-associative operation). For this reason the first section of the chapter is devoted to the definition and basic properties of tournament algebras.

The two tournament solutions known to be computable by binary trees are the Top-Cycle and the Banks' set. Both of them have been mainly studied by political scientists, the motive being that decisions in political assemblies are often reached through successive pairwise votes, the schedule of these votes being specified by the law. This point will be explained in more detail in the section 4 of chapter 10.

8.1. Definition of a Tournament Algebra

An *algebra* $(X, \vee)$ is a pair where X is a set and $\vee$ is an operation on X, that is an application $\vee : X \times X \to X$. Instead of $\vee(x,y)$, one writes $x \vee y$. An algebra is said to be :

- *idempotent* if for all $x \in X$, $x \vee x = x$,
- *commutative* if for all $x, y \in X$, $x \vee y = y \vee x$,
- *associative* if for all $x, y, z \in X$, $x \vee (y \vee z) = (x \vee y) \vee z$.

A tournament naturally defines an algebra :

Definition 8.1.1. The *tournament algebra* associated to a tournament $T \in \mathcal{T}(X)$ is the algebra $(X, \vee_T)$ where, for all $x, y \in X$,

- $x \vee_T x = x$
- if xTy then $x \vee_T y = y \vee_T x = x$.

When there can be no confusion about the tournament under consideration, we simply write $x.y$ or xy instead of $x \vee_T y$. An alternative definition is : $xy = Max(T/\{x,y\})$. Two properties follow from the definition and exactly characterize the tournament algebras.

Proposition 8.1.2. Let $T \in \mathcal{T}(X)$, the tournament algebra $(X, \vee_T)$ is commutative and satisfies

$$\forall (x,y) \in X \times X,\ x \vee_T y \in \{x, y\}.$$

Proposition 8.1.3. Let $(X, \vee)$ be a commutative algebra such that

$$\forall (x,y) \in X \times X,\ x \vee y \in \{x, y\}.$$

Then there exists a unique $T \in \mathcal{T}(X)$ such that $\vee = \vee_T$.

Proof :

Given $(X, \vee)$ one can and must define T by xTy if and only if $x \neq y$ and $x \vee y = x$. ■

Observe also that tournament relations and algebras have the same isomorphisms. Recall that, in an algebra $(X, \vee)$, an element $x \in X$ is *absorbing* if, for all $y \in X$, $x \vee y = y \vee x = x$, and is *neutral* if , for all $y \in X$, $x \vee y = y \vee x = y$. It is straightforward that :

Proposition 8.1.4. Let $x \in X$ and $T \in \mathcal{T}(X)$, x is a Condorcet winner for T if and only if x is absorbing for $(X, \vee_T)$ and x is a Condorcet loser if and only if x is neutral.

Another nice property is that, dealing with tournaments, associativity of the operation is a synonym for transitivity of the relation :

Proposition 8.1.5. A tournament is transitive if and only if its algebra is associative.

Proof :

Let T be transitive and $x, y, z \in X$. For proving that $(xy)z = x(yz)$, the only interesting case is when x, y and z are distinct. Each product is one of the elements x, y and z, and both are equal to y if and only if yTx and yTz. Thus if $(xy)z \neq x(yz)$, xTy or zTx. We can suppose that xTy ; then $(xy)z = xz$. If xTz then $(xy)z = x$ and $x(yz)$ also equals x because xTy and xTz. If zTx, transitivity implies that zTy and $(xy)z = x(yz) = z$. Therefore the algebra is associative.

Conversely, let $x, y, z \in X$ such that xTy and yTz. Clearly $x \neq z$ because a tournament relation is asymmetric, so we just have to prove that $xz = x$. But xTy implies $x = xy$, thus $xz = (xy)z$. By associativity, $xy = x(yz)$ and because yTz, $yz = y$ and $xz = xy = x$. Hence xTz and the relation is transitive. ■

We now introduce a notion of *algebraic application of m variables*. Consider a set Ξ of m symbols, say $\Xi = \{1, ..., m\}$. An *algebraic expression* on Ξ is a formal expression containing symbols and pairs of parenthesis that can be read as a multiple product, the product operation being possibly non-associative. A recursive definition is the following.

Definition 8.1.6. Let Ξ be a finite set,

- each $i \in \Xi$ is an (elementary) algebraic expression on Ξ,
- if f and g are algebraic expressions on Ξ then $(f)(g)$ is an algebraic expression on Ξ.

For instance, $f = [(1)(4)]\{(2)[(3)(4)]\}$ is an algebraic expression on $\{1, 2, 3, 4\}$. One usually does not write the superfluous parentheses and the previous expression is written $f = (14)[2(34)]$. An algebraic expression f is *complete* on Ξ if all the elements of Ξ appear in f. The *order* of f is the number of distinct symbols in f. The *lenght* of f is the number of symbols in f, distinct or not. When needed, the symbols in an algebraic expression f are written as arguments of f, in the example :

$$f(1, 2, 3, 4) = (14)[2(34)],$$

then the algebraic expressions obtained by permutation or substitution of the symbols are easily written, for instance :

$$f(2, 1, 3, 4) = (24)[1(34)]$$

and more generally if f is an algebraic expression on Ξ and π is an application from Ξ to some other set of symbols Ξ', we denote by $f(\pi)$ the algebraic expression on Ξ' obtained by replacing in f each symbol $i \in \Xi$ by $\pi(i) \in \Xi'$.

An algebraic expression f on $\{1, ..., m\}$ with an algebra $(X, \vee)$ define an application $f^\vee$ from X^m to X, given $(x_1, ..., x_m) \in X^m$, $f^\vee(x_1, ..., x_m)$ is computed by substituing the element x_i for the symbol i in f and working out the products according to $\vee$: the *algebraic application* associated to $(X, \vee)$ and f is the application $f^{\vee} : X^m \to X$ defined by :

- If f=i is an elementary expression then $f^{\vee}(x_1, ..., x_m) = x_i$,
- If f=$(f_1)(f_2)$ then $f^{\vee}(x_1, ..., x_m) = f_1{}^{\vee}(x_1, ..., x_m) \vee f_2{}^{\vee}(x_1, ..., x_m)$.

Conversely, an application $F : X^m \to X$ is algebraic for $\vee$ if there exists an algebraic expression such that $F = f^{\vee}$. Remark that there may be several expressions f defining the same algebraic application F. More generally, one can consider any set of symbols Ξ and associate an alternative $a(i) \in X$ to each symbol $i \in \Xi$:

<u>Definition 8.1.7.</u> Let $(X, \vee)$ be an algebra over a finite set X, let f be an algebraic expression on a finite set Ξ of symbols and let a be an application from Ξ to X, the *winner* of $(X, \vee)$ along f according to a is the element $<f, \vee, a>$ of X defined by :

- If f=i is elementary then $<f, \vee, a> = a(i)$.
- If f=$(f_1)(f_2)$ then $<f, \vee, a> = <f_1, \vee, a> \vee <f_2, \vee, a>$.

These ideas are useful for finding tournament winners and defining tournament solutions. If $\vee_T$ is the tournament operation associated to some tournament T, we write $<f, T, a>$ rather than $<f, \vee_T, a>$. Some specific definitions are now introduced, designed for choice in tournaments.

<u>Definition 8.1.8.</u> Let $T \in \mathcal{T}(X)$ be a tournament of order n and Ξ a set of n symbols. A complete algebraic expression on Ξ is called a *board* for X (or for T). A bijection $a : \Xi \to X$ is called a *drawing* for X (or for T).

For any two finite sets Ξ and X, $\mathcal{A}(\Xi, X)$ denotes the set of injections from Ξ to X ; these are bijections if $\#\Xi = \#X = n$, and specifically for $\Xi = \{1, ..., n\}$, we denote by $\mathcal{A}(X)$ the set of bijections from $\{1, ..., n\}$ to X.

In the tournament case, we write f^T, rather than $f^{\vee T}$, the algebraic application defined by *f*. Suppose that *f* is complete with symbols $1, ..., n$, then *f* is a board for tournaments of order *n*. Given such a tournament *T* on a set X of *n* alternatives, a drawing $a \in \mathcal{A}(X)$ assigns an alternative $a(i)$ to each symbol *i* in *f* and the winner of the tournament *T* along the board *f* according to the drawing *a* is the alternative :

$$<f, T, a> = f^T(a(1), ..., a(n)).$$

Example 8.1.9. Consider the algebraic expression *f* previously introduced :

$$f = (14)[2(34)].$$

Let $X = \{x_1, x_2, x_3, x_4\}$ and let $T \in \mathcal{T}(X)$ be defined by

- $x_1 \to x_2, x_3$
- $x_2 \to x_3, x_4$
- $x_3 \to x_4$
- $x_4 \to x_1$.

Take the drawing *a* defined by $a(i) = x_i$, $i=1, ..., 4$. Then

$$<f, T, a> = (x_1\, x_4)(\, x_2(x_3\, x_4)) = x_4(x_2\, x_3) = x_4 x_2 = x_2\, .$$

When the drawing *a* varies among all the possible drawings of $\mathcal{A}(X)$, the board *f* defines a subset of X denoted by $S^f(T)$:

$$S^f(T) = \{\, x \in X : \exists\, a \in \mathcal{A}(X) : x = <f, T, a> \}.$$

Once a board for each possible order *n* is set, one can construct a tournament solution, this is the purpose of the next definition, introduced by two propositions.

Proposition 8.1.10. If *x* is the Condorcet winner of $T \in \mathcal{T}(X)$ and *f* is a board for *T* then for any drawing $a \in \mathcal{A}(X)$, $<f, T, a> = x$.

Proof :

By definition of a drawing there exists i such that $x=a(i)$, and it has been already noticed that a Condorcet winner is absorbing in the tournament algebra. The result follows. ■

<u>Proposition 8.1.11.</u> Let $T \in \mathcal{T}(X)$ and $T' \in \mathcal{T}(X')$ be two isomorphic tournaments, with $\varphi : X \to X'$ a tournament isomorphism, let n be the order of T and T', then a is a drawing for T if and only if $a'=\varphi o a$ is a drawing for T', and moreover, for any board f of order n, $<f, T', a'> = \varphi(<f, T, a>)$.

Proof :

It is obvious that $a \in \mathcal{A}(X)$ iff $a' \in \mathcal{A}(X')$. For any f, $<f, T', a'> = f^{T'}(a'(1), ..., a'(n)) = f^{T'}(\varphi(a(1)), ..., \varphi(a(n)))$, because φ is a tournament morphism, this alternative is also $\varphi(f^{T}(a(1), ..., a(n)))$, that is $\varphi(<f, T, a>)$. ■

<u>Definition 8.1.11.</u> Let $\mathcal{F} = (f_n)_{n \geq 1}$ be a sequence of boards, f_n of order $n \geq 1$; $\mathcal{F}$ is called an *agenda*. For $T \in \mathcal{T}(X)$ of order n, define :

$$S^{\mathcal{F}}(T) = S^{f_n}(T) = \{ x \in X : \exists a \in \mathcal{A}(X) : x = <f_n, T, a> \},$$

then S is a tournament solution called the *algebraic solution* associated to the agenda $\mathcal{F}$. A tournament solution S is algebraic if there exists $\mathcal{F}$ such that $S = S^{\mathcal{F}}$.

8.2. Binary Trees

In the figures 8.1, 8.2 and 8.3, three objects are represented : a *binary tree*, a *labelled binary tree* and a *table of results*. Here are the formal definitions of these objects.

<u>Definition 8.2.1.</u> A binary tree is a digraph, that is a pair (N, A) with N a finite set of *nodes* and $A \subset N \times N$ a finite set of *arcs* such that for $v, w \in N$, $(v, w) \in A$ reads "v is a predecessor of w" or "w is a sucessor of v" and :

- There exists a unique node $w^* \in N$ with no sucessor, w^* is called the *terminal* node.
- Each node exept the terminal node has exactly one sucessor.
- Each node has 0 or 2 predecessors, the nodes with 0 predecessors are called *initial* nodes.

<u>Definition 8.2.2.</u> A labelled binary tree of order n is a triple (N, A, i) where (N, A) is a binary tree and i is an injection from the set N^0 of inital nodes of (N, A) to a set Ξ of n symbols. For $v \in N^0$, $i(v) \in \Xi$ is the *label* of v.

Observe that a labelled binary tree of order n has at least n initial nodes because each label must appear at least once, but it may have more since several initial nodes may have the same label. (If each symbol appears exactly once, this structure is sometimes refered to as a "knockout tournament". Some authors have studied the playing of generalized tournaments on this specific labelled binary trees, see 10.1.4 to 10.1.6.) Labelled binary trees can be used to find winners in a tournament (in fact in any commutative algebra).

<u>Definition 8.2.3.</u> Given a labelled binary tree (N, A, i) of order n with set of symbols Ξ, a tournament $T \in \mathcal{T}(X)$ of order n and a drawing $a \in \mathcal{A}(\Xi, X)$, call *table of results* the application r from N to X defined by :

- If $v \in N$ is an initial node then $r(v) = a(i(v))$.
- If $w \in N$ is not an initial node then $r(w) = r(v)r(v')$, where v and v' are the two predecessors of w.

Then the *winner* of T along (N, A, i) according to a is the alternative $r(w^*)$, where w^* is the terminal node of (N, A).

An orientation of the binary tree (N, A) is a labelling of the set of edges A by the set *{left, right}* such that for any non-initial node w, if v and v' are the two predecessors of w, one of the two edges (v,w) and (v',w) is labelled *left* and the other is labelled *right*. Picture 8.1 shows an oriented binary tree with 5 initial nodes u_1, u_2, u_3, u_4 and u_5. Picture 8.2 shows an oriented labelled binary tree of order 4, obtained with the previous oriented binary tree and the set of symbols $\{1, 2, 3, 4\}$. In picture 8.3 is represented the table of results corresponding to the board of figure 8.2, the tournament T on $X = \{x_1, x_2, x_3, x_4\}$ and the drawing a of example 8.1.8. The alternatives which appear on the nodes of the tree trace the computation of the winner of T along f according to a, where f is the board corresponding to the labelled binary tree of figure 8.2, up to the terminal node w^* which is such that

$$r(w^*) = \langle f, T, a \rangle.$$

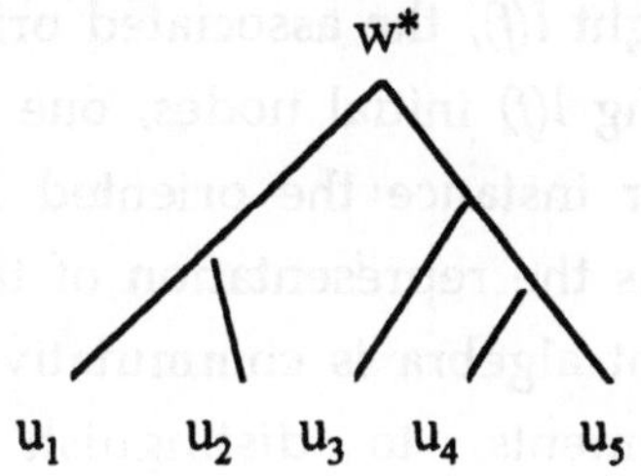

Figure 8.1.

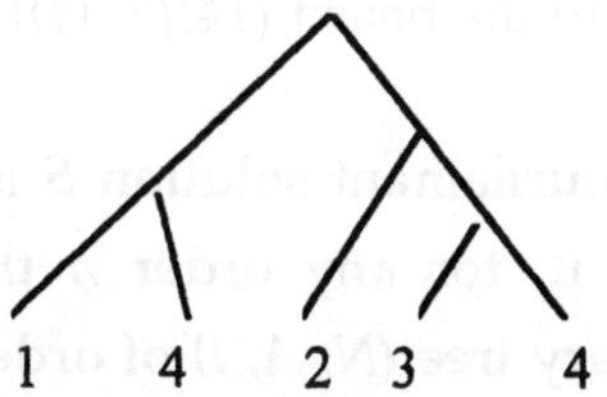

Figure 8.2.

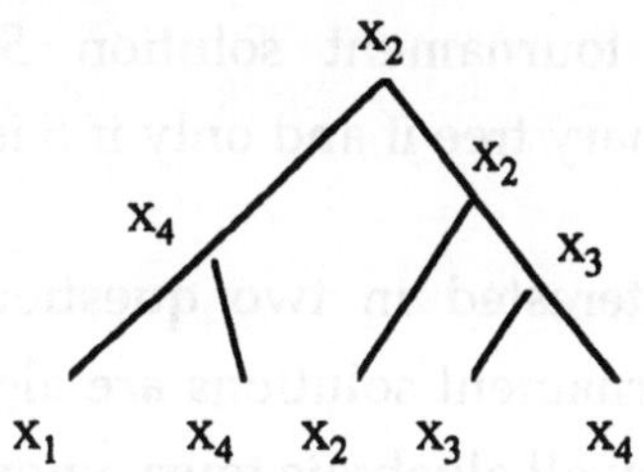

Figure 8.3.

It is not difficult to see that there is a correspondence between oriented labelled binary trees and boards with the same set of symbols : Given an oriented labelled binary tree, start from the terminal node and write one pair of parentheses for each predecessor *v* of the terminal node (one pair on the left and the other on the right), then continue with the left (right) predecessor of *v* in the left (right) parentheses ; write *i(v)* if *v* is an initial node. Conversely,

given a board f of lenght $l(f)$, the associated oriented binary tree is obtained by considering $l(f)$ initial nodes, one non-initial node for each product in f. For instance the oriented labelled binary tree depicted in picture 2 is the representation of the board *(14)[2(34)]*. But since a tournament algebra is commutative, there is no need, dealing with tournaments, to distinguish between different orientations of the same labelled binary tree ; and there is no need to distinguish between boards which only differ by the orientation of their associated labelled binary trees. For instance the board *[2(34)](14)* is equivalent to the board *(14)[2(34)]*.

Definition 8.2.4. A tournament solution S is *computable by binary tree* if, for any order n there exists a labelled binary tree (N, A, i) of order n such that, for any tournament $T \in \mathcal{T}(X)$ of order n, $S(T)$ is the set of winners of T along (N, A, i) for the various drawings of X.

Remark 8.2.5. A tournament solution S is computable by binary tree if and only if S is algebraic.

We shall be interested in two questions : 1) Which ones, among the known tournament solutions are algebraic ? 2) What are the properties shared by all algebraic tournament solutions ?

8.3. An Algebraic Solution : The Top-Cycle

In view of the preceding section, a solution S is algebraic if and only if there exists an agenda such that for any tournament T of order n, the question "Is x a S-winner of T ?" can be answered by seeding the initial nodes of a binary tree of order n through all drawings of X. This requires computation of, at most, $n!$ tables of results. The two main tournament solutions which have been proved to be algebraic are the Top-Cycle and the Banks' set. It is sometimes possible to prove that a given solution is not algebraic, this has been done for the Copeland rule. But in most cases, the question "Is S algebraic?" remains unsolved. In this section we establish a result which was apparently first clearly stated by Miller (1977). The Top-Cycle is algebraic, associated to a sequence of boards called the *simple agenda.*

Definition 8.3.1. Let $(\Phi_n)_{n\geq 1}$ be the sequence of boards defined by

- $\Phi_1 = 1$
- $\forall n \geq 2,\ \Phi_n = (n)(\Phi_{n-1})$,

the board Φ_n has symbols $1, \ldots, n$ and is called the *simple agenda* of order n.

The corresponding labelled binary tree is depicted in figure 8.4. and the simple agenda for 1, 2, ..., 5 alternatives is :

$$\Phi_1 = 1$$
$$\Phi_2 = 2.1$$
$$\Phi_3 = 3(2.1)$$
$$\Phi_4 = 4(3(2.1))$$
$$\Phi_5 = 5(4(3(2.1))).$$

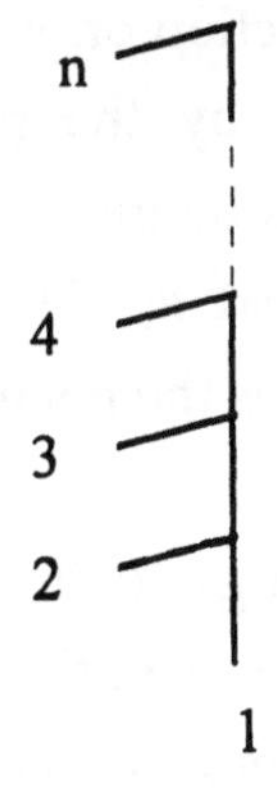

Figure 8.4.

Before proving that the Top-Cycle is the algebraic tournament solution associated with the simple agenda, two lemmas about the Top-Cycle are useful. (Recall that $T\text{-}x$ denotes the restriction of $T \in \mathcal{T}(X)$ to $X\text{-}\{x\}$.)

Lemma 8.3.2. Let $T \in \mathcal{T}(X)$ be a tournament of order at least 2, then : $TC(T) = \{ x \in X : \exists y \in TC(T\text{-}x) : xTy \}$.

Proof :

Let $x \in X$, if y exists such that $y \in TC(T\text{-}x)$ and xTy, y indirectly beats all the elements of $X\text{-}\{x,y\}$ and thus xTy implies $x \in TC(T)$. Conversely, suppose that $x \in TC(T)$ is beaten by all the alternatives of $Y = TC(T\text{-}x)$, then Y is a component of T which beats all the alternatives outside Y, contradicting proposition 1.3.16. ■

Lemma 8.3.3. Let $T \in \mathcal{T}(X)$ be a tournament of order n and $x \in X$, then $x \in TC(T)$ if and only if there exists a drawing $a : \{1, ..., n\} \rightarrow X$ such that $a(n)=x$ and for all $i \in \{1, ..., n\text{-}1\}$, $a(i+1)Ta(i)$.

Proof :

By definition of a drawing all the alternatives appear in the sequence $a(1), ..., a(n)$ thus the existence of a implies $x \in TC(T)$. The converse is obtained by induction on n. For $n=1$ it is trivial ; for $n>1$ consider the restriction $T\text{-}x$, by the previous lemma there exists $y \in TC(T\text{-}x)$ such that xTy, thus by induction there exists a drawing $a : \{1, ..., n\text{-}1\} \rightarrow X\text{-}\{x\}$ such that $a(n\text{-}1)=y$ and for all $i \in \{1, ..., n\text{-}2\}$, $a(i+1)Ta(i)$. Letting $a(n)=x$ gives the result. ■

An alternative x belongs to the Top-Cycle when any other alternative can be reached by a path of some lenght starting at x. Lemma 8.3.3 states that x belongs to the Top-Cycle if and only if x is the first element of a *spanning* (or *hamiltonian*) path in T : only one path is necessary for reaching all the other alternatives.

<u>Theorem 8.3.4.</u> **The Top-Cycle is the algebraic tournament solution associated with the simple agenda.**

Proof :

Let $x \in TC(T)$, it is easy to verify that the drawing a in lemma 8.3.3 is such that $<\Phi_n, T, a> = a(n)=x$. Conversely let $a \in \mathcal{A}(X)$, one has

$$<\Phi_n, T, a> = a(n) <\Phi_{n-1}, T', a'>$$

where T' is T restricted to $X-\{a(n)\}$ and a' is a restricted to $\{1, ..., n-1\}$. The result is then easily seen by induction : $y = <\Phi_{n-1}, T', a'>$ being an element of $TC(T')$, $x=<\Phi_n, T', a>$ is equal to $a(n)$ if $a(n)Ty$ and to y if $yTa(n)$. In the first case $x \in TC(T)$ by lemma 8.3.2 and in the second case $x=y$ beats directly $a(n)$ and indirectly all the other alternatives hence $x \in TC(T)$. ■

An interpretation of the simple agenda is the following : alternatives are ranked according to some drawing a. In the first stage alternative $a(1)$ is considered as a possible best choice. In a second stage, $a(2)$ is proposed for replacing $a(1)$ as a best choice, and the second stage best choice is the best of $a(2)$ and $a(1)$, that is $a(2)a(1)$. The same process then goes on and defines a i-th stage best choice for $i=1, ..., n$. The last of those provisional winners is the winner of the tournament according to the simple agenda and the given drawing. The sequence of "i-th stage best choices" is called the sincere sequence.

<u>Definition 8.3.5.</u> The *sincere sequence* associated with a tournament $T \in \mathcal{T}(X)$ and a drawing $a \in \mathcal{A}(X)$ is the sequence $b = (b_1, ..., b_n)$ of elements of X defined by :

- $b_1 = a(1)$
- $\forall i \in \{1, ..., n-1\}, b_{i+1} = b_i\, a(i+1)$.

The sincere sequence is the one which appears in the table of results of a simple agenda at the non-initial nodes. At the terminal node, $<\Phi_n, T, a> = b_n$. In figure 8.5 is represented the table of results for the tournament used in example 8.1.8., the drawing a such that $a(i)=i$, $i=1, \ldots, 4$, and the simple agenda of order 4.

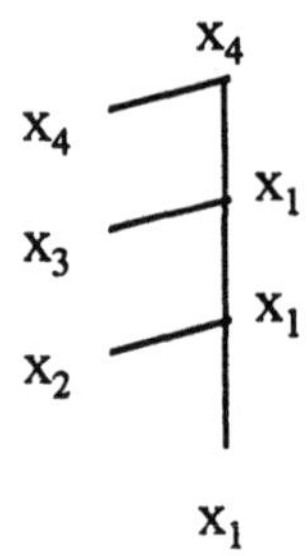

Figure 8.5.

8.4. An Algebraic Solution : The Banks' set

The Banks' set is the algebraic solution obtained with a particular agenda (sequence of boards of all orders) called the *sophisticated* agenda. The sophisticated agenda has been introduced by Miller (1977) and Shepsle and Weingast (1984) to describe the strategic behavior of the voters in front of a simple agenda. They are multistage elimination trees (Moulin 1979) describing sophisticated voting in amendment procedures. For more details on this matter, see Farquharson (1969) and McKelvey and Niemi (1978).

<u>Definition 8.4.1.</u> Let $(\Psi_n)_{n\geq 1}$ be the sequence of boards defined by

- $\Psi_1 = 1$
- $\forall\, n \geq 2,\ \Psi_n\ (1, 2, ..., n) = [\Psi_{n-1}(1,\ \ 3, ..., n)]$ $[\Psi_{n-1}(2, 3, ..., n)]$,

the board Ψ_n has symbols $1, ..., n$ and is called the *sophisticated agenda* of order n.

The corresponding labelled binary tree is depicted in figure 8.6 for the order n=4. The lenght of Ψ_n is 2^{n-1}. The first sophisticated agendas are :

$$\Psi_1 = 1$$
$$\Psi_2 = 12$$
$$\Psi_3 = (13)(23)$$
$$\Psi_4 = [(14)(34)][(24)(34)]$$
$$\Psi_5 = (\ [(15)(45)][(35)(45)]\ )(\ [(25)(45)][(35)(45)]\ ).$$

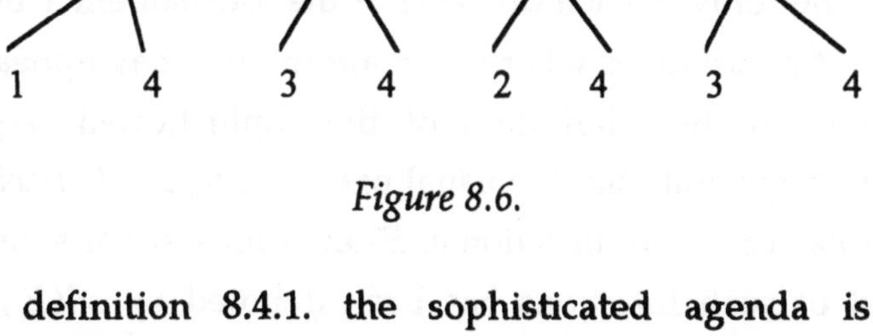

Figure 8.6.

In definition 8.4.1. the sophisticated agenda is defined recursively "from the top": the two subtrees above the terminal node are two sophisticated agendas of order n-1. It is also possible to define the same sequence of boards "from the bottom": knowing Ψ_{n-1} (with symbols $1, 2, ..., n$), Ψ_n is obtained by replacing each initial node of Ψ_{n-1} labelled i by a non-initial node whose two predecessors are initial and labelled i and n. This property is the object of the next proposition.

Proposition 8.4.2. The sophisticated agenda can be defined by :

$$\forall n \geq 2, \Psi_n(1, 2, \ldots, n) = \Psi_{n-1}(1n, 2n, \ldots, (n-1)n).$$

Proof :

By induction on n. There is nothing to prove for $n=1$; suppose that the result holds up to $n>1$, then $\Psi_n(1, 2, \ldots, n)$ is equal to :

$$[\Psi_{n-1}(1, 3, \ldots, n)]\ [\Psi_{n-1}(2, 3, \ldots, n)]$$

by definition. Now by induction this writes :

$$[\Psi_{n-1}(1(n+1), 3(n+1), \ldots, n(n+1))]\ [\Psi_{n-1}(2(n+1), 3(n+1), \ldots, n(n+1))]$$

and thus again by the definition :

$$\Psi_n(1(n+1), 2(n+1), \ldots, n(n+1)).$$

■

The winner along a sophisticated agenda can also be computed with the help of a recursively defined sequence called the sophisticated sequence. In the sequel, we will have to associate such a sequence not only to each drawing of the tournament T but also to sequences of alternatives where each alternative may appear several times. Therefore our definition of the sophisticated sequence is slightly more general than the usual one. Call *repeated drawing* of the set of alternatives X an injection $a: \Xi \to X$ from a set of symbols Ξ to X. The set of such injections has been denoted by $\mathcal{A}(\Xi, X)$. In a repeated drawing each alternatives appears at least once but may appear several times. If Ξ has $m \geq n$ symbols and f is a board of order m then $<f, T, a>$ is well defined.

<u>Definition 8.4.3.</u> Let $m \geq n$, $a:\{1, ..., m\} \rightarrow X$ a repeated drawing of X and $T \in \mathcal{T}(X)$. The *sophisticated sequence* associated to a and T is the sequence $(\beta_1, \beta_2, ..., \beta_m) \in X^m$ defined by $\beta_m = a(m)$ and for $i \in \{1, ..., m-1\}$,

- if $\forall j \in \{i+1, ..., m\}$, $a(i)T\beta_j$ then $\beta_i = a(i)$,
- if not, then $\beta_i = \beta_{i+1}$.

<u>Theorem 8.4.4.</u> Let $a:\{1, ..., m\} \rightarrow X$ be a repeated drawing of X and $T \in \mathcal{T}(X)$, then $< \Psi_m, T, a> = \beta_1$,where β is the sophisticated agenda associated with a and T.

Proof :

By induction on m. For $m=1$ it must be the case that $\#X=n=1$ and the result is true. Let a be a repeated drawing of order m for some set X with $\#X=n \leq m$ and let $T \in \mathcal{T}(X)$. By the proposition,

$$\Psi_m(a(1), ..., a(m-1), a(m)) = \Psi_{m-1}(a'(1), ..., a'(m-1))$$

with $a'(i) = a(i)a(m)$. The application $a' : \{1, ..., m-1\} \rightarrow X$ is a repeated drawing of order $m-1$ on its image $Y \subset X$. Denote by $\beta_m, \beta_{m-1}, ... \beta_1$ and $\beta'_{m-1}, ..., \beta'_1$ the sophisticated sequences associated to a and a'. We claim that :

$$\forall i \in \{1, ..., m-1\}, \beta_i = \beta'_i.$$

Since $\beta_m = a(m)$ and $\beta_{m-1} = a(m-1)a(m)$ it is clear that $\beta'_{m-1} = \beta_{m-1}$. Suppose that $\beta_j = \beta'_j$ is true for $j=m-1, ..., i+1$; by definition of the sophisticated sequences :

$$\beta_i = \beta_{i+1}, \text{ or } \beta_i = a(i) \text{ if } a(i)\beta_j = a(i) \text{ for all } j = m, m-1, ..., i+1$$

$$\beta'_i = \beta'_{i+1}, \text{ or } \beta'_i = a(i) \text{ if } a(i)\beta'_j = a(i) \text{ for all } j = m-1, ..., i+1.$$

Obviously, if $\beta_i = \beta_{i+1}$ and $\beta'_i = \beta'_{i+1}$ then $\beta'_i = \beta_i$. Suppose $a(i)\beta_j = a(i)$ for all $j = m, m-1, ..., i+1$, then for $j=m$, $a(i)a(m)=a(i)$ and for $j \in \{m-1, ..., i+1\}$, $[a(i)a(m)]\, \beta'_j = a(i)\, \beta_j = a(i)$ thus $\beta'_i = \beta_i = a(i)$. Suppose that $a(i)\beta'_j = a(i)$ for all $j = m\text{-}1, ..., i+1$, then two cases must be considered :

- If $a(i)a(m) = a(i)$ then for $j \in \{m\text{-}1, ..., i+1\}$, $a(i)\ \beta_j = a(i)$ and *for* $j=m$, $a(m)\ \beta_m = a(m)$ therefore $\beta_i = a(i) = \beta'_i$.
- If $a(i)a(m) = a(m)$ then for $j \in \{m\text{-}1, ..., i+1\}$, $a(m)\ \beta_j = a(m)$ and this implies that $a(m) = \beta_m = \beta_{m-1} = ... = \beta_{i+1}$; then $\beta_i = \beta'_i = a(m)$.

One thus finds that $\beta'_i = \beta_i$ and proves the claim. Hence $\beta_1 = \beta'_1$ and the induction hypothesis applies :

$$\beta'_1 = \Psi_{m-1}(a'(1), ..., a'(m-1))$$

and give the result :

$$< \Psi_m, T, a > = \Psi_m(a(1), ..., a(m-1), a(m)) = \beta'_1 .$$

■

The result holds in the particuliar case $m=n$ so that one obtains :

Corollary 8.4.5. The winner of a tournament T of order n along the sophisticated agenda of order n for a drawing a is the first term (β_1) of the sophisticated sequence associated to T and a.

This result is usually proven (Shepsle and Weingast 1984, Moulin 1986, Reid 1991a and 1991b) by reducing the board of order n "by the top" using definition 8.4.1. rather than "by the bottom" using proposition 8.4.2.

Theorem 8.4.6. Let $T \in \mathcal{T}(X)$ and $m \geq n = \#X$ then :
$\{ < \Psi_m, T, a > : a \in \mathcal{A}(\{1, ..., m\}, X) \} = B(T)$.

Proof :

Let $x \in B(T)$, there exists a transitive maximal chain $x=x_1\ T\ x_2\ ...\ T\ x_k$ with x on top. Let $a : \{1, ..., m\} \to X$ be a repeated drawing such that $a(i)=x_{n-i+1}$ for $i=m, ..., m\text{-}k+1$, it is easily seen that for the associated sophisticated sequence, $\beta_i = x$ for all i, thus $x = <\Psi_m, T, a>$.

Conversely let a be such that $x = <\Psi_m, T, a>$, then for the associated sophisticated sequence, $\beta_1 = x$. By definition each new term in the sequence β beats all the previous ones hence the set $Y = \{\beta_m, \beta_{m-1}, ..., \beta_1\}$ forms a transitive chain with x on top. If there exists $z \in X$ such that zTY then, because $z=a(i)$ for some i, z beats $\beta_m, \beta_{m-1}, ..., \beta_{i+1}$ implies $z = \beta_i$, a contradiction. Hence Y is maximal and $x \in B(T)$. ■

By letting $m=n$ in the previous theorem one obtains the main result of this section.

<u>Theorem 8.4.7</u> The Banks'set is the algebraic solution associated with the sophisticated agenda.

8.5. Properties of Algebraic Solutions

This section contains some results, positive and negative, concerning the problem "What can be said about algebraic solutions in general?" As the reader will observe, not much is known about this problem. Two positive results (8.5.1 and 8.5.2) known about the properties of algebraic solutions are due respectively to McKelvey and Niemi (1978) and to Moulin (1986).

<u>Theorem 8.5.1</u> Any algebraic tournament solution selects in the Top-Cycle.

Proof :

Let $\varphi : X \rightarrow \{0, 1\}$ be defined by $\varphi(x) = 1$ if $x \in TC(T)$ and $\varphi(x) = 0$ if not, φ is a tournament morphism from T to the tournament on $\{0,1\}$ such that $1.0 = 1$. For any drawing a of X, $\varphi o a$ is a repeated drawing of $\{0,1\}$. For a board f,

$$\varphi(<f, T, a>) = <f, \varphi[T], \varphi o a> = 1$$

hence $<f, T, a> \in TC(T)$. ■

Observe that the proof given here is just a re-writing of the one that H. Moulin put in some more poetic terms ("Paint the outcomes of $TC(T)$ in red..."). Unfortunately this theorem is of little help for deciding whether a given solution is or is not algebraic, because all the interesting tournament solutions are refinements of the Top-Cycle. Moreover it has been noticed that the Top-Cycle is itself algebraic, so that one cannot hope to strentghten the theorem by replacing the Top-Cycle by some other finer tournament solution. The next theorem is more informative.

<u>Theorem 8.5.2.</u> Any algebraic tournament solution is wealy composition-consistent.

Proof :

Following the definition of weak composition-consistency, consider T and T' two tournaments on the same set X of alternatives such that T and T' only differ on some subset Y of X. Denote $X^* = X - Y \cup \{Y\}$, that is replace in X the subset Y by a single element. Let $\varphi : X \rightarrow X^*$ be defined by $\varphi(x)=x$ if $x \notin Y$ and $\varphi(x)=Y$ if $x \in Y$. Then φ is a tournament morphism for both T and T' with the same image $T^* = \varphi[T] = \varphi[T']$. Hence for any board f and drawing a, $\varphi(<T, f, a>) = \varphi(<T', f, a>)$ and the result follows. ■

With this theorem one can prove that there exist tournament solutions which are *not* algebraic (Moulin 1986, Dutta and Sen 1993) because they are not weakly composition-consistent :

Corollary 8.5.3. The Copeland and Markov tournament solutions are not algebraic.

We now come to the monotonicity properties of the algebraic solutions. The first remark is due to Moulin (1986), it does not exactly concerns algebraic solutions but the solving of tournaments by binary trees with a fixed drawing.

Example 8.5.4. Consider the board of order 4 : $f = 3(4(2(13)))$ depicted in figure 8.7 (observe that this board is not a simple agenda because the label 3 appears twice) and the tournament T on $\{x_1, x_2, x_3, x_4\}$ used in example 8.1.8. Let a be the drawing $a(i)= x_i$ for $i=1, ..., 4$. Then $<f, T, a> = x_3$. Now let T' be identical to T exept that $x_3 \rightarrow x_1$ instead of $x_1 \rightarrow x_3$ one finds $<f, T', a> = x_2$.

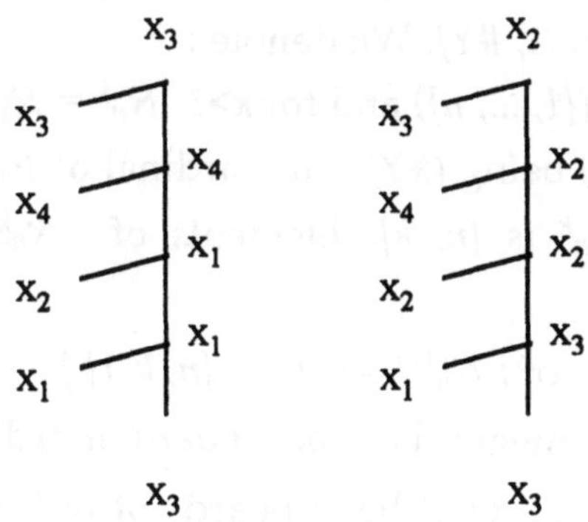

Figure 8.7.

This example is paradoxal because strengthening the winning alternative makes it lose. It reveals that, the drawing a being fixed, the application $T \rightarrow <f, T, a>$ from $\mathcal{T}(X)$ to X lacks monotonicity. But this application does not define a tournament solution because it is not neutral (the names of the alternatives matter). It will be proven that there exists an algebraic solution which is not monotonous, according to the definition (2.3.1) of Monotonicity for tournament solutions. It is not difficult to see that the choice correspondence $\mathcal{S}f$ defined on the tournaments of order 4 by the board $f = 4(3(2(14)))$ of the previous example coincides with Top-Cycle and is therefore monotonous according to definition 2.3.1. Conversely, it is obvious that if, for every drawing a, the choice function $T \rightarrow <f, T, a>$ is monotonous then the choice correspondence

$$\mathcal{S}f : T \rightarrow \{<f, T, a> : a \in \mathcal{A}(X)\}$$

is monotonous.

The following construction is based on an iteration of the sophisticated agenda (Coughlan and LeBreton, 1994). Some notations are required. For any positive integer n, the iterated factorials of n are recursively defined by :

$$[n, 0] = n \text{ and for } k>0,\ [n, k] = [n, k-1]!.$$

For any finite set Y, $\aleph(Y)$ is the set of orderings of Y, that is the set of all bijections $Y \rightarrow \{1, ..., \#Y\}$. We denote :

$$\aleph_n^1 = \aleph(\{1, ..., n\}) \text{ and for } k>1,\ \aleph_n^k = \aleph(\aleph_n^{k-1}).$$

The cardinal of $\aleph(Y)$ being $(\#Y)!$, the cardinal of $\aleph_n^1$ is $n! = [n, 1]$ and the cardinal of $\aleph_n^k$ is $[n, k]$. Elements of $\aleph_n^k$ will be denoted by σ^k :

$$\sigma^k : \aleph_n^{k-1} \rightarrow \{1, ..., [n, k-1]\}$$

so that $\sigma^k(\sigma^{k-1})$ is an integer. Let f be a board of order m with symbols $1, ..., m$ and let $g(1), ..., g(m)$ be m boards of order n, with the same symbols $1, ..., n$, then $f(g)$ is the board obtained by substituing in f each i by $(g(i))$. The symbols in $f(g)$ are $1, ..., n$. The number n being fixed, we shall define by induction on k an application F_n^k which associates to each $\sigma^k \in \aleph_n^k$ a board $F_n^k(\sigma^k)$ with symbols $1, ..., n$.

<u>Definition 8.5.5.</u> For $\sigma^k \in \aleph_n^k$, the *extension* by σ^k of the sophisticated agenda of order n is the board of order n denoted $F_n^k(\sigma^k)$ inductively defined by :

- For k=1 and $\sigma^1 \in \aleph_n^1 : F_n^1(\sigma^1) = \Psi_n(\sigma^1)$.
- For k>1 and $\sigma^k \in \aleph_n^k : F_n^k(\sigma^k) = \Psi_{[n,\ k-1]}(g)$, where g is defined by:
  $\forall \sigma^{k-1} \in \aleph_n^{k-1},\ g(\sigma^k(\sigma^{k-1})) = F_n^{k-1}(\sigma^{k-1})$.

<u>Proposition 8.5.6.</u> Let $T \in \mathcal{T}(X)$ be a tournament of order n and let a be a drawing of X with symbols $\{1, \ldots, n\}$, then for any $k \geq 1$:

$$\{ < F_n^k(\sigma^k), T, a > : \sigma^k \in \aleph_n^k \} = B^k(T).$$

Proof :

By induction on k. For k=1, each $\sigma^1 \in \aleph_n^1$ is just a permutation of $\{1, \ldots, n\}$ and $<F_n^1(\sigma^1), T, a> = <\Psi_n, T, a o \sigma^1>$, but clearly, $\{a o \sigma^1 : \sigma^1 \in \aleph_n^1\} = \mathcal{A}(\{1, \ldots, n\}, X)$ hence the result for k=1 by Theorem 8.4.6. For k>1, for any $\sigma^{k-1} \in \aleph_n^{k-1}$, by induction hypothesis,

$$\{ < F_n^{k-1}(\sigma^{k-1}), T, a > : \sigma^{k-1} \in \aleph_n^{k-1} \} = B^{k-1}(T).$$

Let $\sigma^k \in \aleph_n^k$. For any $i \in \{1, \ldots, [n, k-1]\}$ there exists a unique $\sigma^{k-1} \in \aleph_n^{k-1}$ such that $i = \sigma^k(\sigma^{k-1})$, let $a'(i) = < F_n^{k-1}(\sigma^{k-1}), T, a >$, the induction hypothesis means that a' is a repeated drawing of $B^{k-1}(T)$. Therefore Theorem 8.4.6. implies that the alternative $<F_n^k(\sigma^k), T, a>$ is in $B^k(T)$. This proves :

$$\{ < F_n^k(\sigma^k), T, a > : \sigma^k \in \aleph_n^k \} \subset B^k(T).$$

Let $x \in B^k(T) = B(B^{k-1}(T))$, denote $n' = \#B^{k-1}(T)$ and $T' = T/B^{k-1}(T)$, there exists $a'' \in \mathcal{A}(\{1, \ldots, n'\}, B^{k-1}(T))$ such that $x = <\Psi_{n'}, T', a''>$. One can build a repeated drawing a' of $B^{k-1}(T)$ which still has x as a winner and which is compatible with the construction. First observe that x is the winner for any a' of the form :

$$a''(1), \ldots, a''(1), a''(2), \ldots, a''(k-1), a''(k), \ldots, a''(k), a''(k+1), \ldots, a''(n')$$

obtained by repeating some elements in the sequence a'' (this observation is obvious in view of 8.3.3. and 8.4.4.). Next observe that, when σ^k varies in $\aleph_n{}^k$, the alternatives of $B^{k-1}(T)$ appear as $< F_n{}^{k-1}(\sigma^{k-1}), T, a >$ in all possible orders, under the only constraint that the number of times each one appears is fixed : this is because the board for B^{k-1} is repeated $[n, k-1]$ times, using all $\aleph_n{}^{k-1}$. It follows that there exists an ordering $\sigma^k \in \aleph_n{}^k$ and a repeated drawing $a' \in \mathcal{A}([n, k-1], B^{k-1}(T))$ such that $x = < \Psi_{[n,\, k-1]}, T', a'>$ and for all $i \in \{1, ..., [n, k-1]\}$, $a'(i) = < F_n{}^{k-1}(\sigma^{k-1}), T, a>$ for σ^{k-1} such that $i = \sigma^k(\sigma^{k-1})$. Then :

$$<F_n{}^k(\sigma^k), T, a> = < \Psi_{[n,\, k-1]}, T', a'> = x$$

and the proof is complete. ■

This proposition implies that it is possible to construct an algebraic tournament solution finer that B^k, for any k. For each $n \geq 1$ choose a $\sigma^k \in \aleph_n{}^k$ and for any tournament T on X of order n let $BB^k(T) = \{ < F_n{}^k(\sigma^k), T, a > : a \in \mathcal{A}(X) \}$, then BB^k is an algebraic solution and $BB^k(T) \subset B^k(T)$. Since for T of order n, $B^n(T) = B^\infty(T)$, it is also possible to construct an algebraic solution finer than B^∞ : for any n, choose $\sigma^n \in \aleph_n{}^n$ and let $BB^\infty(T) = \{ < F_n{}^n(\sigma^n), T, a > : a \in \mathcal{A}(X) \}$,. One may conjecture that $BB^\infty(T) = B^\infty(T)$, but this has not been proven. Nevertheless the inclusion $BB^\infty \subset B^\infty$ is sufficient for the next proposition. One preliminary lemma about the iterated Banks' set is needed.

<u>Lemma 8.5.7.</u> For $k \geq 3$, including $k=\infty$, the Iterated Banks' set, B^k, is not monotonous.

Proof :

The counter-example is the same as for the Iterated Uncovered set (Theorem 5.2.4.). The set of alternatives is $X = \{1, 2, 3, 1', 2', 3'\}$, the tournament T_1 is depicted in figure 5.3. The two transitive chains $(1, 2, 3')$ and $(1', 1, 3)$ are maximal thus $1 \in B(T_1)$

and $1' \in B(T_1)$. By symmetry $B(T_1)=X$ and thus for any k, $B^k(T_1)=X$. Let $T'_1 = T_{1<1',2'>}$, it has been noticed that for $k \geq 3$, $UC^k(T'_1) = UC^3(T'_1) = \{1,2,3\}$. Since by proposition 7.1.8. $B^3 \subset UC^3$ it is easy to find that $B^k(T'_1) = B^3(T'_1) = \{1,2,3\}$, hence $1'$ is no longer a winner and B^k is not monotonous. ■

Proposition 8.5.8. There exists a non monotonous algebraic tournament solution.

Proof :

Consider the tournaments T_1 and T'_1 used in the proofs of 5.2.4. and 8.5.7. Since $1' \in B^3(T_1)$, proposition 8.5.6. implies that there exists $\sigma^3 \in \aleph_6^3$ such that

$$1' \in \{ < F_6^3(\sigma^3), T_1, a > : a \in \mathcal{A}(X) \}.$$

For T'_1, proposition 8.5.6. implies that

$$\{ < F_6^3(\sigma^3), T'_1, a > : a \in \mathcal{A}(X) \} \subset B^3(T'_1)$$

and thus

$$1' \notin \{ < F_6^3(\sigma^3), T'_1, a > : a \in \mathcal{A}(X) \}.$$

As a consequence, any algebraic tournament solution such that its board of order 6 is $F_6^3(\sigma^3)$ fails to be monotonous.

■

This proposition is surprising, and stronger than the example by Hervé Moulin showing that, a drawing being fixed, strenghtening a winner can make him or her lose (cf. example 8.5.4 and figure 8.7). The crucial point in Moulin's example is that x_3, by not defeating x_1 at an early stage in the agenda, allows x_1 to later defeat those other candidates that x_3 would not beat (x_2 in the example). The paradox therefore essentially rests on the order of appearance of the alternatives (the drawing). Proposition 5.3.8 is quite different. Say that a competitor *has a chance to win* if there exists a drawing that will make him or her win. The proposition says that the following situation may arise : Altough other competitors, including a certain y, are stronger than you, you have a chance to win providing the drawing will be favorable to you. But if you were to be stronger than y, then there could be no favorable drawing for you.

Of course the construction above has a purely theoretical interest because the extented sophisticated agenda is a very long expression, the lenght of $F_n^k(\sigma^k)$ is $2^{[n,k-1]-1}$. For instance the binary tree used as a counter-example, with k=3 and n=6, has $2^{720!}$ - 1 nodes, a large number.

9 - Copeland Value of a Solution

In the previous chapters, several ideas have been introduced, allowing to define solution concepts. It has been possible to show that these solutions were really different ones, but we did not tackled the problem of knowing whether they are *very* different. In order to quantify the gap between two solutions S and S', the most natural idea is, for a given tournament T, to compute the number of alternatives on which they disagree, that is the cardinal of the symmetric difference $S(T) \Delta S(T')$. This would enable to compute disagreement indices for correspondences S and S'. But we will not follow this line of reasoning because these computations would have few meaning concerning composition-consistent tournament solutions. So we drop the idea of quantifying the disagreement between any two solutions and we simply try to compare any solution S to the Copeland one.

9.1. Definition of the Copeland Value

A tournament solution S will be considered being close to the Copeland one if, for any tournament T, $S(T)$ contains points whose Copeland score is not very different from the maximal Copeland score in T. So we state the following definition.

Definition 9.1.1.: Let S be a tournament solution and T be a tournament on X, the *Copeland Value of S on T*, $CV_S(T)$ is :

$$CV_S(T) = \frac{\sup\{s(x), x \in S(T)\}}{\sup\{s(x), x \in X\}}.$$

Clearly, $0 < CV_S(T) \leq 1$, and $CV_S(T) = 1$ if and only if $S(T) \cap C(T) \neq \emptyset$; moreover $S(T) \subset S'(T)$ implies $CV_S(T) \leq CV_{S'}(T)$.

Definition 9.1.2. The *Copeland Value* of a solution S, CV_S is :

$$CV_S = \inf\{CV_S(T) : T \in \mathcal{T}\}.$$

Again, one has : $0 \leq CV_S \leq 1$ and $S \subset S' \Rightarrow CV_S \leq CV_{S'}$.

9.2. Computation of Some Copeland Values

We now give either the exact value or an upper bound for the Copeland Values of the main tournament solutions. Obviously, the Copeland Value of the Copeland solution is equal to 1. Because the Copeland solution is finer than the Top-Cycle and the Uncovered Set, one directly gets the following theorem:

<u>Theorem 9.2.1.:</u> $CV_{TC} = CV_{UC} = CV_C = 1$.

Other solutions have Copeland Values strictly smaller than 1.

<u>Proposition 9.2.2.:</u> $CV_{TCoUC} \leq 1/2$.

Proof:

We construct a family T_n of tournaments of order $(2n+1)\,(n+1)+1$ in the following way: The set of alternatives X_n is $X_n = \{x_o\} \cup X^1 \cup X^2$, with :

- $X^1 = \{x_1^i : i=1,\ldots\ldots,2n+1\}$,
- $X^2 = \bigcup_{i=1}^{2n+1} X^{2,i}$ with $\#X^{2,i} = n$.

The relation T_n is defined by :

- $X^1 \rightarrow x_o$,
- $x_o \rightarrow X^2$,
- $\forall i \in \{1,\ldots,2n+1\}$, $x_1^i \rightarrow X^{2,i}$,
- The restriction T'_n of T_n to $X^1 \cup X^2$ is the product of the cyclone of order $2n+1$, C_{2n+1} by the $2n+1$ sub-tournaments $T_n/\{x_1^i\} \cup X^{2,i}$, for $i=1,\ldots,2n+1$.
- $T_n/X^{2,i}$ is arbitrary chosen.

The figure 9.1. represents the tournament T_1, of order 7. It is easily checked that $UC(T_n) = X^1 \cup \{x_0\}$ (x_1^i covers the points of $X^{2,i}$) and that $TCoUC\,(T_n) = X^1$ (x_o is a Condorcet looser on $X^1 \cup \{x_0\}$). Counting the scores gives

$s(x_o) = n(2n+1)$,

$s(x_1^i) = 1+n+n(n+1) = n^2 + 2n + 1$ for all i, and

$s(x^2) \leq n(n+1) + (n-1)$ for all $x^2 \in X^2$.

As a consequence, for $n \geq 2$ the unique Copeland winner of T_n is x_0, and the Copeland Value of T_n is $CV_{TCoUC}\,(T_n) = \dfrac{n^2+2n+1}{2n^2+n}$, which tends to $1/2$ when n tends to infinity. Hence the result. ■

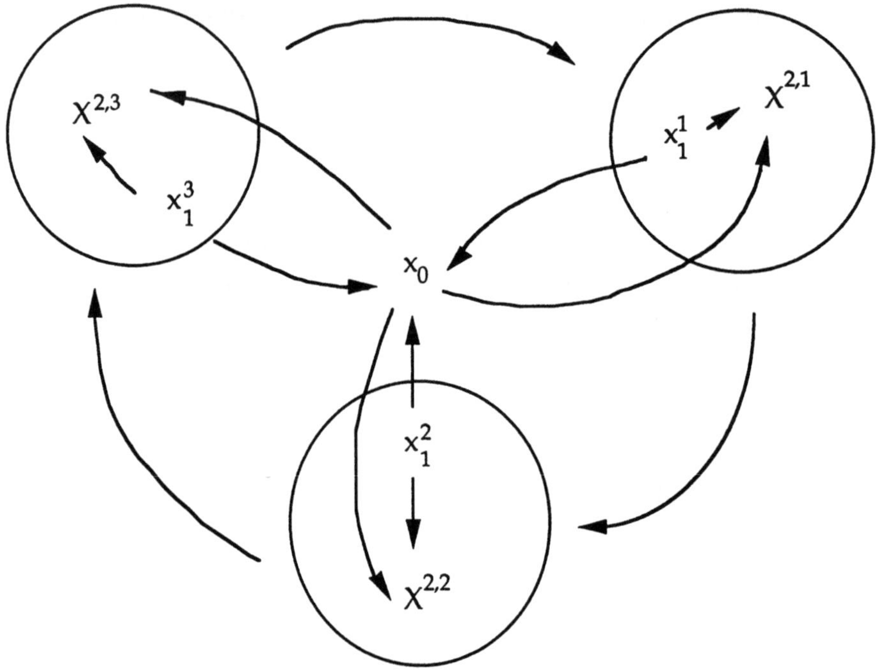

Figure 9.1.

Proposition 9.2.3: $CV_{BP} \geq 1/2$

Proof :

Let T be a tournament defined on X, denote by p the equilibrium strategy for the tournament game $J(T)$ and $Y = BP(T) = Supp(p)$. Also denote $Z = X\text{-}Y$. For $y \in Y$, let $u(y) = \#\{z \in Z : zTy\}$, and $v(y) = \#\{x \in Y : yTx\}$, we shall need two lemmas.

Lemma 9.2.4.: $$\sum_{y \in Y} p(y)u(y) \leq \frac{\#Z}{2}.$$

Proof :

This sum can be written: $\sum_{y\in Y}\sum_{\substack{z\in Z\\ zTy}} p(y) = \sum_{z\in Z}\sum_{\substack{y\in Y\\ zTy}} p(y).$

According to 6.2.3.(ii), $\sum_{\substack{y\in Y\\ zTy}} p(y) \leq 1/2$ for all $z \in Z$ hence the lemma. ■

Lemma 9.2.5.: $\sum_{y\in Y} p(y)\, s(y) \geq \frac{\#Y-1}{2} + \frac{\#Z}{2}.$

Proof :

Recall that $s(y)$ is the Copeland score of y, thus $s(y) = v(y) + (\#Z - u(y))$. One has:

$$\sum_{y\in Y} p(y)\, v(y) = \sum_{y\in Y}\sum_{\substack{x\in Y\\ yTx}} p(y) = \sum_{x\in Y} \Big(\sum_{\substack{y\in Y\\ yTx}} p(y)\Big).$$

But it is known from 6.2.3.(i) that for $x \in Y$,

$$\sum_{\substack{y\in Y\\ yTx}} p(y) = \sum_{\substack{y\in Y\\ xTy}} p(y) = \frac{1-p(x)}{2},$$

as a consequence:

$$\sum_{y\in Y} p(y)v(y) = \frac{1}{2}\sum_{x\in Y} (1-p(x)) = \frac{\#Y-1}{2}.$$

Moreover, from the preceding lemma,

$$\sum_{y\in Y} (\#Z-u(y))p(y) = \#Z\sum_{y\in Y} p(y) - \sum_{y\in Y} u(y)p(y) \geq \frac{\#Z}{2},$$

which concludes the proof of lemma 9.2.5. ■

End of the proof of the proposition : From lemma 9.2.5 it follows in particular that there exists $y_0 \in Y$ such that $s(y_0) \geq \frac{\#Y+\#Z-1}{2}$. If the Copeland winner of T is in Y, $CV_{BP}(T) = 1$, if not, it is in Z, but any point in Z being dominated by at least one point in Y one has for any $z \in Z$, $s(z) \leq \#Y + \#Z - 2$ and so $s(y_0) \geq \frac{1}{2}s(z)$ and $CV_{BP}(T) \geq 1/2$, which completes the demonstration of the proposition 9.3.

■

The two previous propositions allow to find the Copeland Value of the solutions which are finer than *TCoUC* and coarser than *BP*.

Theorem 9.2.6.: For $S = TCoUC, UC^2, \dots UC^K, \dots UC^\infty, MC$ and BP, $CV_S = 1/2$.

Proof:

It has already been noticed that if $S \subset S'$, then $CV_S \le CV_{S'}$, the theorem follows from the inclusions $BP \subset MC \subset UC^\infty \subset \dots \subset UC^K \subset \dots \subset UC^2 \subset TCoUC$ and from the preceding propositions. ■

It is also possible to find the Copeland Value of the Slater solution, this value is again equal to 1/2.

Theorem 9.2.7.: $CV_{SL} = 1/2$.

Proof:

It has been noticed (proposition 5.1.12) that *SL* is finer than *TCoUC*, therefore $CV_{SL} \le 1/2$, and it is also known (proposition 3.4.3) that in a tournament T of order n, for any $x \in SL(T)$, $s(x) \ge En[\frac{n}{2}]$ (the integer part of $n/2$). One deduces that $CV_{SL} \ge 1/2$ and the theorem. ■

Nevertheless there are solution concepts whose Copeland Values are strictly smaller than *1/2*. Such is the case for the Banks solution (and thus for the Tournament Equilibrium Set). Unfortunately, the exact value of this indicator is not known for these solutions. We are only able to give an upper bound: *1/3*.

Proposition 9.2.8.: $CV_{TEQ} \leq CV_B \leq 1/3$

Sketch of the proof :

The first inequality is obvious. For the second one we again need the construction of a sequence of counter-examples. A preliminary lemma is of some interest : for a tournament T, $o(T)$ denotes the order of the tournament and $l(T)$ the maximum length of a transitive chain of T, that is to say the maximal order of the transitive sub-tournaments of T.

Lemma 9.2.9.: For all $\varepsilon > 0$, there exists a regular tournament T such that $\frac{l(T)}{o(T)} \leq \varepsilon$.

Proof:

Let T_1 be a cyclical tournament of order 5, one has $l(T_1) = 3$, for any $k \geq 1$ define inductively T_{k+1} as the composition-product of T_k by $o(T_k)$ instances of itself:

$$T_{k+1} = \Pi(T_k ; T_k, \ldots, T_k) = T_k \otimes T_k,$$

then it is easily seen that $o(T_{k+1}) = [o(T_k)]^2$, that T_{k+1} is regular and that $l(T_{k+1}) = [l(T_k)]^2$, so for any k: $o(T_k) = 5^{2^{k-1}}$ and $l(T_k) = 3^{2^{k-1}}$. Since $(3/5)^{2^{k-1}}$ tends to 0 when k tends to infinity, the lemma is proved. ■

End of the proof of the proposition : The idea of the proof is identical to the one that has been used for example 7.1.4. : We consider a tournament whose chains are of relatively small length, we add another tournament defined with these chains as vertices, then a third tournament, dominating the first one and being dominated by the second one. This construction being rather long, we refer the reader to the original article (Laffond, Laslier and LeBreton 1994b)

■

10 - From Tournaments to Choice and Voting

This final chapter offers a short survey on possible generalizations and applications of some of the concepts developed for tournaments. We do not consider the case of incomplete tournaments, where some comparisons are missing, but the case of so-called "generalized tournaments", where the result of each comparison is quantitative rather than qualitative. A good example of a generalized tournament is found in Social Choice Theory when two alternatives are compared according to the number of individuals who prefer the one or the other. Another example is the case where one knows for each x and y the probability that x beats y. Particular applications of tournaments (and generalized tournaments) are given in two models in Voting Theory: voting with mediators and voting with agendas. These applications explain why some specific tournament solutions have been defined and studied.

10.1. Generalized Tournaments

Several concepts developed about tournaments can be extended to richer structures. We consider the following structure, called a generalized tournament. We give the definition using matrix terminology, but it is clear that other equivalent definitions can be given.

Definition 10.1.1.: A real square matrix M of size $n \times n$ is a *generalized tournament* matrix if there exists $m > 0$ such that for all $x \in \{1, ..., n\}$ and all $y \in \{1, ..., n\}$ with $y \neq x$,

$$M_{xx} = 0$$
$$M_{xy} + M_{yx} = m.$$

A generalized tournament is a comparison structure such that, unlike in a tournament, comparisons are quantitative rather than qualitative but, like in a tournament:

(i) Comparing an element x with itself is meaningless: the assumption that $M_{xx}=0$ is a matter of convention, and we would develop the same ideas letting $M_{xx} = \frac{m}{2}$.

(ii) There is a kind of anti-symmetry in the pairwise comparisons: x is good with respect to y (M_{xy} large) as much as y is bad with respect to x (M_{yx} small). Of course, tournaments are generalized tournaments such that $m = 1$ and all the M_{xy} are 0 or 1. We denote $X = \{1, ..., n\}$.

The vector b defined by:

$$b_x = \sum_{y \in X} M_{xy}$$

is a generalization of the vector of Copeland scores. We shall refer to b as the vector of *Borda scores*. Compared to the Copeland score, the

Borda score is a very rich concept. In a Social Choice Theory framework, when one interprets M_{xy} as the number of individuals who prefer x to y, the Borda score has a simple but remarkable property: according to its definition, the Borda score of an alternative x is the sum over all the alternatives y of the votes that x gets when opposed to y. As the reader will easily check, the Borda score of x is also the sum over all the individuals i of the number of alternatives y such that i prefers x to y. As such, the Borda score is a particular case of the positional scoring rules studied by Saary (1994). The Borda score is particularily appealing in a context where the set of voters, rather than the set of alternatives may vary, and can be axiomatized in this context. On the problem of voting with a variable electorate and the axiomatization of the Borda rule, see Smith (1973), Young (1974, 1975) and Myerson (1995). There is no reason for the Condorcet winner, when it exists, to have a large Borda score, and choosing the "Borda winner" is not a majoritarian procedure.

An important literature is devoted to the analysis of ratio-scaled comparison matrices, that is positive square matrices R such that $R_{xy}R_{yx} = 1$. Curiously, this literature has essentially used linear techniques (such as eigenvalue methods analog to the "Long Path method" presented in chapter 3, section 2) for this non-linear structure. If the pairwise comparisons are ordinal then one writes the ratio-scaled comparison matrix R with $R_{xy} \in \{a, 1, 1/a\}$ for some constant a; then the results of the linear analysis of R depend on the value of the *ad hoc* parameter a. The interested reader should read one of the many books of T. L. Saaty on the "Analytical Hierarchy Process", for instance Saaty (1980), the book *The Analytic Hierarchy Process: Applications and study* (Golden, Wasil and Harker 1989), Vargas and Whittaver (1990) or Genest, Lapointe and Drury (1993). Other references on this popular method and related others are

Saaty (1977), Johnson, Beine and Wang (1979), Jensen (1984, 1986), Aupetit and Genest (1993) and Genest and Rivest (1994).

The Markov method can also be defined for generalized tournaments. It is then called the *Self-consistent Rule* (SCR) and has been studied by Levchenkov (1995a, b) and the other authors mentionned in 3.3. The equations for finding the *self-consistent scores*, $scr(x)$ are:

$$scr(x) = scr(x)\sum_{y\in X}\frac{M_{xy}}{(n-1)m} + \sum_{y\in X}\frac{M_{yx}}{(n-1)m}scr(y).$$

The Slater method translates into the so-called *Kemeny Rule*. If U is a ranking of X, the quantity:

$$\Delta(U, M) = \Sigma\{M_{xy} : yUx\}$$

is interpreted as the disagreement between U and M. A ranking U such that this disagreement is minimum is called a *Kemeny order* for M. (Kemeny 1959, Kemeny and Snell 1962, Young and Levenglick 1978, Monjardet 1990). The Kemeny rule is a majoritarian procedure: it chooses the Condorcet winner whenever it exists. Young (1988) argues that this rule, among the various majoritarian rules now defined, remains true to Condorcet's thought.

The Multivariate Descriptions presented in chapter 4 are also valid for generalized tournaments (Laslier 1996a); one has to replace "Copeland score" by "Borda score" and the same results hold, including the fact that the correlation between the Borda score and most of the principal components is zero (proposition 4.1.9.).

To a generalized tournament M, one associates a two-player symmetric zero-sum game (X, g) by letting:

$$g(x, y) = M_{xy} - M_{yx}.$$

But, unless some other assumptions on M are made, any two-player zero sum game has this form. As a consequence only the general results on two-player, zero-sum games are true, and the theory developed about the Uncovered set, the Minimal Covering set and the Bipartisan set does not carry over immediately to the case of generalized tournaments In particular the notion of covering is no longer identical to weak dominance. As explained in section 5 of chapter 6, the theory does carry over when some additional restrictions are imposed on the class of admissible generalized tournaments. By chance, these restrictions make sense in the theory of voting (see the next section) and allow for the definition of a generalized Bipartisan set.

It seems that the contestation process has no appealing generalization for the class of generalized tournaments. One can imagine generalizing these concepts to the case of binary relations which are not complete (and maybe: which are not asymmetric). Schwartz (1990) makes propositions in this direction.

The question of playing generalized tournaments on binary trees has been adressed by several scholars and has led to sometime suprising results. If M is a positive generalized tournament matrix with $M_{ij} + M_{ji} = 1$, one can interpret M_{ij} as the probability that i beats j. Given a board, f, of size n and a drawing, a, of $\{1, ...,n\}$, one can compute the probability that alternative i wins the generalized tournament M. Let $w_i(M, f, a)$ denote this probability. If the drawing is choosen at random among the $n!$ possible drawings, the probability that i wins M along f is the mathematical expectation of $w_i(M, f, a)$ and is denoted by $w_i(M, f)$. It turns out that questions related to the monotonicity properties of $w_i(M, f)$ are often uneasy to answer. The results obtained concern special classes of boards and generalized tournaments. We mention these results without proof.

Definition 10.1.2. : A board such that each label appears exactly once at the terminal nodes is called a *knock-out* board.

Knock-out boards (Hartigan 1966) are usual direct elimination schedules in sport competitions. A simple agenda is a knock-out board but a sophisticated agenda is not. Another familiar example is the *balanced* knock-out board, where 2^m players are planned to play m rounds, half of them being eliminated at each round (m-th of final, ..., semifinal, final). For each order n, there is a finite number of knock-out boards.

There seems to be no problem for ranking the alternatives from best ($i=1$) to worse ($i=n$) if M satisfies the following condition (*cf.* David 1963).

Definition 10.1.3. : A positive square matrix M of size n *is strongly stochasticly transitive* if for all $i, j \in \{1, ..., n\}$, $M_{ij} + M_{ji} = 1$ and :

$$i<j \text{ implies that } \forall k \in \{1, ..., n\}, M_{ik} > M_{jk}.$$

Observe that for such a matrix, $M_{ii} = 1/2$ and $M_{ij} > 1/2$ iff $i<j$.

Proposition 10.1.4. (Chung and Hwang 1978) : Suppose that the strongly stochastically transitive matrix M is such that there exist numbers h_i ($i =1, ..., n$) with $M_{ij} = h_i/(h_i + h_j)$. Then for any knock-out board f,

$$i<j \text{ implies } w_i(M, f) \geq w_j(M, f).$$

The matrix M such that $M_{ij} = h_i/(h_i + h_j)$ is called *a Bradley-Terry preference scheme* (Bradley and Terry 1952). The previous proposition still holds for *diluted* Bradley-Terry scheme of the form

$$M_{ij} = (ah_i + (1-a)h_j)/(h_i + h_j) \text{ for } 1/2 < a < 1,$$

as proved in Hwang, Zongzhen and Yao (1991). Without a specified scheme, the same result can be proved to hold on average:

Proposition 10.1.5. (Hwang and Hsuan 1980) : Suppose that M is strongly stochastically transitive and let $w_i(M)$ denote the average of $w_i(M, f)$ over all possible knock-out boards, then

$$i<j \text{ implies } w_i(M) > w_j(M).$$

Chung and Hwang (1978) made the natural conjecture that if M is strongly stochasticly transitive then $i<j$ implies $w_i(M, f) \geq w_j(M, f)$ for any knock-out board. However this conjecture was shown to be false by R. Israel, who provided an amazing counter-example.

Proposition 10.1.6. (Israel 1982) : There exists a strongly stochasticly transitive matrix M of order 17 (on $X = \{1, ..., 17\}$) and a knock-out board f such that $w_{17}(M, f) > w_{16}(M, f)$.

Israel's example is counter-intuitive because the hypothesis of strong stochastic transitivity seems to leave no room for any kind of intransitivity or rank-reversal. When directly confronted the one to the other, *16* wins more often than it looses against *17* and for any other alternative they eventually meet along the binary tree, *16* also does better than *17*. It is therefore strange that *17* wins on the all more often with a random drawing. Hwang (1982) showed that the paradox cannot involve the most likely winning alternative : under strong stochastic transitivity, $w_1(M,f) \geq w_i(M,f)$ and Chen and Hwang (1988) showed that the paradox can neither take place in a balanced knock-out board.

10.2. Social Choice

In Social Choice Theory, one considers a finite set I of *individuals* and a finite set X of *alternatives*. Individuals have transitive preferences P_i over X and

$$P = (P_i)_{i \in I} \in [OC(X)]^I$$

is called a preference *profile* (definition 2.1.1.). An alternative $x \in X$ is *unanimously preferred* to another alternative y if for any individual i, xP_iy. Suppose that all the individual preferences are strict, then x is unanimously preferred to $y \neq x$ if all the individuals strictly prefer x to y.

Definition 10.2.1.: Let $P \in [OS(X)]^I$ and $x, y \in X$. We say that x *Pareto-dominates* y if $\forall i \in I, xP_iy$.
The set of alternatives which are not Pareto-dominated is called the *Pareto-set* of P and is denoted $PAR(P)$.

Clearly, $PAR(P)$ is never empty. Since a Pareto-dominated alternative can be unanimously improved upon, it is a natural requirement that a social choice correspondence only selects Pareto-undominated alternatives.

Suppose that the sets X and I are finite and that $\#I$ is odd. Then the majority relation $M(P)$ defined by:

$$x\, M(P)\, y \Leftrightarrow \#\{i \in I : xP_iy\} > \#\{i \in I : yP_ix\}$$

is a tournament on X. Given a tournament solution S one can consider the social choice correspondence $S' = SoM$ defined by $S'(P) = S(M(P))$. For instance Merlin and Saari (1995a,b) provide a thorough discussion of a typical tournament concept (the Copeland method) considered as a social choice correspondence. Usually the social choice correspondence S' receives the same name as the tournament solution S, for instance, one calls "Uncovered set of P" the Uncovered set of the majority tournament $\mathcal{M}(P)$. Thus, it is worth asking the question of the Pareto-optimality of tournament solutions.

Definition 10.2.2.: A tournament solution S is *Pareto-consistent* if for any profile $P \in [\mathcal{OS}(X)]^I$, $S(\mathcal{M}(P)) \subset PAR(P)$.

Clearly if S_1 is finer than S_2 and S_2 is Pareto-consistent then S_1 is also Pareto-consistent. The following theorem will be sufficient to decide for most tournament solutions whether they are Pareto-consistent or not.

Theorem 10.2.3.: (i) The Top-Cycle is not Pareto-consistent.
(ii) The Uncovered set is Pareto-consistent.

Proof :

(i) Let $X = \{1, 1', 2, 3\}$ and $I = \{i, j, k\}$. Let $P \in [\mathcal{OS}(X)]^I$ be defined by:

$1 \, P_i \, 1' \, P_i \, 2 \, P_i \, 3$

$2 \, P_j \, 3 \, P_j \, 1 \, P_j \, 1'$

$3 \, P_k \, 1 \, P_k \, 1' \, P_k \, 2.$

Then it is easy to see that 1 Pareto dominates $1'$ and that no other alternative is Pareto-dominated, hence :

$$PAR(P) = \{1, 2, 3\}.$$

But for the tournament $\mathcal{M}(P)$, the cycle $1 \to 2 \to 3 \to 1' \to 2 \to 3 \to 1$ shows that :

$$TC(\mathcal{M}(P)) = \{1, 1', 2, 3, 4\}.$$

(ii) We just have to prove that x covers y if x Pareto-dominates y. Suppose that x Pareto dominates y, then clearly $x \, \mathcal{M}(P) \, y$. Let $z \in X$ such that $y \, \mathcal{M}(P) \, z$, for all the individuals i such that $yP_i \, z$, transitivity of P_i implies that $xP_i \, z$. Since these individuals form a majority, $x \, \mathcal{M}(P) \, z$, and it follows that x covers y. ■

The fact that the Top-Cycle is not Pareto-consistent is, for social scientists, a strong motivation for searching for other majoritarian choice correspondences. The fact that the Uncovered set is Pareto-consistent implies that the other tournament solutions presented in this book also define Pareto-consistent social choice correspondences.

The tournament *M(P)* used in the proof of this theorem is the same one that we used for proving that the Top-Cycle is not composition-consistent (proposition 2.4.8.). It is possible to generalize the notions of composition-product, composition-consistency and composition-consistent hull to the case of preference profiles and social choice correspondences (Laffond, Lainé, and Laslier 1996). It turns out that scoring rules are not composition-consistent, for instance the composition-consistent hull of the Borda rule is the Pareto choice correspondence. This is not surprising since the Borda rule satisfies a consistency axiom at the level of the voters rather than at the level of the alternatives. In fact there exists no Paretian composition-consistent social choice correspondence based (like scoring rules) on the ranks of the alternatives in the individuals' preferences (Laslier 1996c). The Kemeny rule is not composition-consistent but the generalized Bipartisan set is composition-consistent and can be axiomatized like in Theorem 6.3.10 once the right concepts of Borda-regularity and Borda-dominance have been defined for the generalized case, to replace Copeland-regularity and Copeland-dominance.

10.3. Voting with Mediators

Given the preference profile $P = (P_i)_{i \in I} \in [\mathcal{OS}(X)]^I$ with I the set of individuals and X the set of alternatives, the *size of the majority* in favor of x against y is:

$$M_{xy} = \#\{i \in I : xP_i y\}$$

and the majority relation $\mathcal{M}(P)$ is: $x\ \mathcal{M}(P)\ y$ iff $M_{xy} \geq M_{yx}$.

Suppose that individual preferences are strict and suppose that $\#I$ is odd. Then, as we mentioned, $\mathcal{M}(P)$ is a tournament. But the generalized tournament M also has some interesting properties, for any $x, y \in X$ with $x \neq y$,

- $M_{xy} + M_{yx} = \#I$ is an odd integer
- $M_{xy} \neq M_{yx}$ (because one is odd and the other even).

Therefore the associated zero-sum game (X, g), with $g(x, y) = M_{xy} - M_{yx}$ have the properties that for $x \neq y$:

- $g(x, y)$ is an odd integer
- $g(x, y) \neq 0$.

It turns out that these properties are sufficient to guarantee existence and unicity of the Weak Saddle (theorem 6.5.1.) and of the mixed Nash equilibrium (theorem 6.5.2.) just like in the tournament case. Therefore, it is worth interpreting and comparing the two symmetric two-player zero-sum games defined by a (strict, odd) preference profile.

<u>**Definition 10.3.1.:**</u> Let $P \in [\mathcal{OS}(X)]^I$ be a strict preference profile with an odd number, $\#I$, of individual. Denote, for $x, y \in X$

$$g^P(x, y) = M_{xy} - M_{yx} \text{ and}$$
$$g^m(x, y) = sgn(g^P(x, y)).$$

Then (X, g^P) and (X, g^m) are respectively called the *plurality* and *majority* games of P.

One interpretation of these games is that the two "players" are two mediators who propose alternatives to the individuals. Then each voter votes for one mediator or the other according to her preferences on the proposed alternatives. If these mediators are interpreted as political parties, then one obtains a model of a pure electoral competition among "downsian" parties, that is to say purely competitive symmetric parties with no *a priori* ideological positions (Laffond, Laslier and LeBreton 1994a). This kind of model is standard in political theory, see for instance Downs (1957) and Ordeshook (1986).

In the plurality game, the number of votes obtained by a party matters: it is better to win the election by a large margin. In the majority game the only objective is to win and the size of majority does not matter. Are these two games really different? It is easy to see that the two games are identical with respect to pure Nash equilibria.

<u>Remark 10.3.2.:</u> The plurality game (X, g^P) of P has a pure Nash equilibrium if and only if the majority game (X, g^m) has one. For both games, the equilibrium strategy is to play the Condorcet winner of $\mathcal{M}(P)$, and if there is no Condorcet winner, there is no pure Nash equilibrium.

But in many interesting cases, no Condorcet winner exists for the majority tournament $\mathcal{M}(P)$. Considering the domination relations in those games, one finds :

<u>Proposition 10.3.3.:</u> If x dominates y for the plurality game, then x dominates y for the majority game (that is: x covers y in the majority tournament).

Proof :

By definition 6.1.6. applied to the symmetric game (X, g^P), x dominates y means that for any $z \in X$:

$$M_{xz} - M_{zx} \geq M_{yz} - M_{zy}.$$

This implies the same inequalities for $g^m(x, z)$ and $g^m(y, z)$ and the result. ■

The converse of this proposition does not hold, but at least if x covers y for the majority game, then it can not be the case that y covers x for the plurality game. A corollary of proposition 10.3.3. is :

<u>Proposition 10.3.4.:</u> The Weak Saddle of the majority game (that is: the Minimal Covering set of the majority tournament) is included in the Weak Saddle of the plurality game.

Proof:

Let Y denote the Weak Saddle of the plurality game. For any $z \in X - Y$, z is a dominated strategy for the plurality game restricted to $Y \cup \{z\}$. By proposition 10.2.3., z is thus dominated for the majority game restricted to $Y \cup \{z\}$. Hence, Y is a covering set for $M(P)$ and the result follows. ■

In the general case, the tournament Minimal Covering set is a *strict* subset of the plurality Weak Saddle.

For the games played in mixed strategies, we know that each of the majority mixed games and the plurality mixed games have a unique equilibrium. Call generalized Bipartisan set, or *plurality-Bipartisan* set the support of the Nash equilibrium of the plurality game and, by contrast, call *majority-Bipartisan* set the Bipartisan set of the majority tournament. Laffond, Laslier and LeBreton (1994a) gave the example of a profile for which these two sets are disjoint. Observe that because the support of the Nash equilibrium is included in the Weak Saddle, proposition 10.3.4. implies that both Bipartisan sets are subsets of the plurality Weak Saddle.

The mixed plurality game can receive a political interpretation. Recall that the payoff for this game is

$$g^P(p, q) = \sum_{x,y \in x} p(x)q(y)(M_{xy} - M_{yx})$$

for any $p, q \in \Delta_X$.

Consider now the following model. Each individual $i \in I$ in the population identifies party A with an alternative $a(i) \in X$ and party B with an alternative $b(i) \in X$. If $a(i)\ P_i\ b(i)$ then i votes for A, if $b(i)\ P_i\ a(i)$ then i votes for B and if $a(i) = b(i)$ then i chooses at random who to vote for.

Suppose that, for each i, $a(i)$ and $b(i)$ are two random variables on X, $a(i)$ and $b(i)$ being independent:

$$Pr[a(i) = c \text{ and } b(i) = y] = Pr[a(i) = x] \times Pr[b(i) = y] = p(i, x)\, q(i, y).$$

Then the expected number of votes in favor of A is:

$$\mathrm{E}[N_A] = \sum_{i \in I} \sum_{x \in X} \sum_{y \in X} p(i,x) q(i,y) \varepsilon(i,x,y).$$

with $\varepsilon(i, x, y) =$ $+1$ if $x P_i y$
0 if $y P_i x$
$½$ if $x = y$.

Suppose that for all i, $p(i, c) = p(x)$ and $q(i, x) = q(x)$ ("homogeneous identification hypothesis"), then

$$\mathrm{E}[N_A] = \sum_{x \neq y} p(x)q(y)M_{xy} + \frac{1}{2}\sum_x p(x)q(x)$$

$$\mathrm{E}[N_A] = \frac{1}{2} g^P(p,q) + \frac{\#I}{2}.$$

and similarly

$$E[N_B]=\frac{1}{2}g^P(q,p)+\frac{\#I}{2}.$$

This means that, up to an affine transformation which is of no real importance, the mixed game $g^P(p, q)$ describes the competition between two parties under the hypothesis that:

- The identification by an individual of a party to an alternative is homogenous in the population of voters.
- For each individual, the identification of the two parties to two alternatives is made independently.
- Parties' variables of action are the global parameters (probability distributions) of the identification party/alternative.
- Parties' objectives are the expected numbers of votes.

This model can be extended to the case of several parties playing a "Borda-competition game" (Laffond, Laslier, LeBreton, 1995b). The mixed majority game has received no convincing political interpretation.

10.4. Voting with Agendas

One very common procedure by which groups of people reach a decision is by successive elimination of alternatives. The simple agenda, and more general binary trees, describe such procedures: an agenda-setter decides on a drawing of the set of alternatives, in a simple agenda the drawing defines the order in which the various possible alternatives will be presented to the individuals and each pairwise comparison is performed on the basis of majority voting.

It has been early noticed (see Black (1958) and McLean and Urken (1995) for historical references) that if voters vote *sincerely*, that is to say if their votes always express their true preferences, then the agenda-setter has, in the absence of a Condorcet winner in the set of possible alternatives, a very important power. Indeed, it follows from McGarvey's theorem (proposition 2.1.3.) and the study of the algebraic solution associated to the simple agenda (theorem 8.3.5.) that the agenda-setter has the power to have the final decision being any alternative in the Top-Cycle of the majority tournament.

This observation calls for a study of procedural decision schemes other than the simple agenda and for the study of possible strategic behavior on the side of the voters. The study of other procedural decision schemes, for instance other types of boards than the simple agendas (cf. Ordeshook and Schwartz 1987) seems to be difficult, as can be deduced from the small number of results in our chapter 8, section 5. An important literature, starting with Farquharson (1969), has been devoted to the strategic behavior of voters facing a binary procedure such as a simple agenda, see for instance Miller (1977), McKelvey and Niemi (1978) or Moulin (1979) after Gibbard (1973) and Satterthwaite (1975).

To motivate the discussion, consider for instance a simple agenda over three alternatives x, y, z presented in the order $a(1) = x$, $a(2) = y$, $a(3) = z$. Suppose that for the majority tournament, yTx and zTy. Then sincere voting leads to the choice of z. But if xTz, those voters who prefer x to z and y to x should better vote, at the first stage, for x rather than y in order that, at the second and final stage, the alternatives under consideration be $\{z, x\}$ rather than $\{z, y\}$.

Therefore, intelligent voters can have an incentive to misrepresent their preferences. It turns out that the game which models these strategic misrepresentations of preferences possess a particular equilibrium, called the *sophisticated equilibrium*, obtained by iterative elimination of weakly dominated strategy (Moulin 1979, 1983). The game is *dominance solvable*. Consequently one usually considers that the sophisticated equilibrium is a good descriptive concept for modelling strategic voting.

With the tool of sophisticated equilibria, scholars have studied the power of the agenda-setter in the case of the simple agenda. The lesson of this literature is that the power of the agenda-setter, although it does not vanish, is *less* important with strategic behavior than it is in sincere voting. With a simple agenda, it is possible to precisely describe this power: the set of possible outcomes is exactly the Banks' set (Schepsle and Weingast 1984, Banks 1985). We now explain this result.

Let $a \in \mathcal{A}(X)$ be a drawing of X. According to the simple agenda, alternatives will be presented in the order $a(1), a(2), a(3), \ldots$ The initial choice between $a(1)$ and $a(2)$ is, for each voter acting strategically, the choice between two consequences: either they will be facing the next agenda $a_{-2} = a(1), a(3), \ldots, a(n)$ on $X - \{a(2)\}$, or they will be facing the agenda $a_{-1} = a(2), a(3), \ldots, a(n)$ on $X - \{a(1)\}$. Therefore, if $f_n(a)$ is the outcome of the strategic voting on a one gets that they vote for $a(1)$ or $a(2)$ according to their preferences between $f_{n-1}(a_{-1})$ and $f_{n-1}(a_{-2})$. The majority wins so that the final outcome is, using the algebraic notations of Chapter 8 :

$$f_n(a) = f_{n-1}(a_{-1})\, f_{n-1}(a_{-2}).$$

One recognizes the recursive definition (8.4.1) of the sophisticated agendas.

Theorem 10.4.1.: Strategic voting on a simple agenda results in choosing the winner of the associated sophisticated agenda.

Proof :

The main argument has been given above. For a detailed proof and more on the sophisticated equilibrium, see Shepsle and Weingast (1984), Reid (1991a), Myerson (1991) or Ordeshook (1992).

■

As a consequence, introducing strategic behavior on the side of the voters reduces the power of the agenda-setter: the set of possible outcomes shrinks from the Top-Cycle of the majority tournament to its Banks' set. In particular, it follows from theorem 10.2.3. and the inclusion $B \subset UC$ that the introduction of strategic behavior rules out the eventuality of choosing Pareto-dominated outcomes when a simple agenda is in use.

Annex - Summary Tables

A.1. Relations between the Main Solutions

Table n°1

	TC	*UC*	UC^{∞}	*MC*	*BP*	*B*	*TEQ*	*C*	*SL*
TC									
UC	$\subset$								
UC^{∞}	$\subset$	$\subset$							
MC	$\subset$	$\subset$	$\subset$						
BP	$\subset$	$\subset$	$\subset$	$\subset$					
B	$\subset$	$\subset$	$\cap$	$\cap$	(a)				
TEQ	$\subset$	$\subset$	$\subset$	(b)	(a)	$\subset$			
C	$\subset$	$\subset$	Ø	Ø	Ø	Ø	Ø		
SL	$\subset$	$\subset$	Ø	Ø	Ø	Ø	Ø	Ø	

(a) According to the propositions 7.1.9 and 7.1.10, there exists $T \in \mathcal{T}_{29}$ such that $B(T) \subset BP(T)$ and $B(T) \neq BP(T)$, and there exists $T' \in \mathcal{T}_6$ such that $BP(T') \subset B(T')$ and $B(T') \neq BP(T')$. It is not known whether the intersection of *B* and *BP* can be empty. Same remarks for *TEQ* and *BP*.

(b) The inclusion $TEQ \subset MC$ remains a conjecture.

This table reads: at the intersection of line S_1 and column S_2,

"$\subset$" means : $\forall T \in \mathcal{T}, S_1(T) \subset S_2(T)$ and $\exists T \in \mathcal{T} : S_1(T) \neq S_2(T)$

"Ø" means : $\exists T \in \mathcal{T}, S_1(T) \cap S_2(T) = \emptyset$,

"$\cap$" means : $\forall T \in \mathcal{T}, S_1(T) \cap S_2(T) \neq \emptyset$.

A.2. Properties of the Main Solutions

Table n°2

	TC	*UC*	UC^{∞}	*MC*	*BP*	*B*	*TEQ*	*SL*	*C*
Monotonicity	yes	yes	no	yes	yes	yes	?	yes	yes
Independence of the losers	yes	no	no	yes	yes	no	?	no	no
Idempotency	yes	no	yes	yes	yes	no	?	no	no
Aïzerman Property	yes	yes	no	yes	yes	yes	?	no	no
Strong Superset Property	yes	no	no	yes	yes	no	?	no	no
Composition-Consistency	no	yes	yes	yes	yes	yes	yes	no	no
Weak Comp.-Consist.	yes	yes	yes	yes	yes	yes	yes	yes	no
Regularity	yes	yes	yes	yes	yes	no	no	yes	yes
Copeland Value	1	1	1/2	1/2	1/2	≤1/3	≤1/3	1/2	1

A.3. Games and Tournaments Concepts

Table n°3

Game (X, g)	Tournament T
pure strategy	alternative
$g(x, y) = +1$	x beats y (xTy)
pure Nash equilibrium	Condorcet winner
x dominates y	x covers y
undominated strategies	Uncovered set $UC(T)$
sequentially undominated strat.	Iterated Uncovered set $UC^{\infty}(T)$
Shapley's weak saddle	Minimal Covering set $MC(T)$
support of the mixed equilibrium	Bipartisan set $BP(T)$
?	Minimal Gain score $\zeta(x)$

A.4. An Example

Consider the tournament *T* on *X* = *{1, 2, 3, 4, 5, 6, 7}* analized in chapter 4 and taken as an example in 6.4.13 and 7.2.7. The tournament is depicted in figure 4.1, multivariate descriptions are given in figures 4.2, 4.3 and 4.4

- *X* = *{1, 2, 3, 4, 5, 6, 7}* is the Top-Cycle.
- *{1, 2, 4}* is the Uncovered set, Iterated Uncovered set, Minimal Covering set, Bipartisan set, Bank's set, Tournament Equilibrium set.
- *{1, 2}* is the Weak Uncovered set *à la* Laffond and Lainé (2 beats 4 and all the points that 4 beats exept one).
- *{2, 4}* is the Weak Uncovered set *à la* Levchenkov (2 beats all the points that *1* beats plus *4* and *7*).
- *2* is the unique winner for the Copeland, Long-Path and Markov scores
- *1* is the unique winner for Slater and the minimal gain score.

Observe that there are three Slater orders: *(1, 2, 3, 4, 5, 6, 7)*, *(1, 2, 3, 5, 6, 4, 7)* and *(1, 2, 3, 6, 4, 5, 7)*. These orders only differ by the ranking of the 3-cycle *{4, 5, 6}*. The Slater method clearly nominates *1* as the winner, *2* in second position, *3* in third and *7* in last, and indicates that *4*, *5* and *6* are *ex aequo* and next to last.

The following table summarizes the results, it includes the scores along the first axis of inertia for both the complete euclidean description (distance d_2) and the multidimensional scaling (distance d_1). It also includes the mixed Nash equilibrium although these probabilities are not to be interpreted as scores.

Table n°4

	Copeland score	Long-Path score	Markov score	axis 1 dist. d_2	axis 1 dist. d_1	Nash proba	Min Gain score	TEQ score
1	4	.533	.561	1.623	1.921	1/3	$+\infty$	1/3
2	5	.547	.738	1.971	2.591	1/3	2	1/3
3	4	.391	.246	1.118	1.464	0	1/2	0
4	3	.366	.241	-1.395	-1.100	1/3	2	1/3
5	2	.176	.050	-0.228	-0.617	0	0	0
6	2	.230	.009	-0.733	-0.906	0	1/2	0
7	1	.212	.112	-2.356	-3.353	0	1/2	0

Index of Main Notations

Sets

$\varnothing$	The empty-set
$X \cup Y$	Union of X and Y
$X \cap Y$	Intersection of X and Y
$X-Y$	Set-difference of X and Y
$X \subset Y$	X is included in Y, in the large sens, $X=Y$ implies $X \subset Y$
$\#X$	Cardinal of X, sometimes denoted $Card(X)$
$\mathcal{P}(X)$	Set of subsets of X

Numbers

$\mathbb{N}$	Positive integers
$\mathbb{N}^*$	Strictly positive integers
$\mathbb{R}$	Real numbers
$\mathbb{C}$	Complex numbers
Z_n	integers modulo n
$En[x]$	Integer part of the real number x
$Supp(p)$	Support of the probability p
Δ_X	Simplex on X

Functions

$\mathcal{A}(X, Y)$	Set of injections from X to Y.
$\mathcal{A}(X)$	Set of bijections from $\{1, ..., \#X\}$ to X ("drawings')
$\aleph_n{}^k$	Set of k-iterated enumerations of $\{1, ..., n\}$ (8.5)

$\sigma(X)$	Permutation group of X

Linear algebra

$\mathbf{1}$	Collumn vector with 1 in each row
$\mathbf{I}$	Identity matrix
$\mathbf{J}$	Square matrix with 1 in all entries
M^T	Transpose of the matrix (or vector) M
$<x, y>$	Scalar product of vectors x and y
$\|\|x\|\|$	Euclidean norm of the vector x

Binary relations

$\mathcal{OC}(X)$	Set of complete orderings of X
$\mathcal{OS}(X)$	Set of linear orders on X
$Max(R)$	Set of maximal elements for R
$\mathcal{R}(X)$	Set of binary relations on X
$\mathcal{T}(X)$	Set of tournaments on X
$\mathcal{T}$	Class of finite tournaments
$\mathcal{T}_n$	Class of tournaments of order n
$TS(R)$	Top-set of R (1.5.10)

Tournaments, general

$Aut(T)$	Group of automorphisms of T (1.1)
$\mathcal{C}(T)$	Set of covering sets for T (5.3.1)
$Cond(T)$	Set of Condorcet winners of T (2.2.2)
$o(T)$	Order of T (1.2)
$T^+(x)$	Successors of x (1.4)
$T^-(x)$	Predecessors of x (1.4)
T/Y	Restriction of T to Y (1.1)
$T\text{-}x$	Restriction of T to $X\text{-}\{x\}$ (2.3)
$T<x,y>$	Tournament obtained by reversing the arrow between x and y (2.3)
xTy	x beats y for the tournament T (1.1)
$x \rightarrow y$	x beats y, for a given tournament (1.1)

$x \to Y$	x beats all the elements of Y (1.1)
$x \Rightarrow y$	x covers y (5.1.1)
$\Delta(T, T')$	Slater distance from T to T' (3.4.1)
Π	Multiple composition product (1.3.5)
$\otimes$	Product (1.3.6)
$\vee_T$	Operation associated to T (8.1)

Tournaments, specific

$<_n$	Usual linear ordering of $\{1, ..., n\}$
C_n	Cyclone of order n (1.2.1)
Q_n	Quadratic residus tournament of order n (1.2.2)

Scores

$k^u(x)$	Score of x along the u-th axis of inertia (4.1)
$lp(x)$	Long-path score of x (3.2.1)
$\bar{p}(x)$	Markov score of x (3.3.1)
$s(x)$	Copeland score of x (outscore) (1.4.2)
$s^+(x)$	Copeland score of x (outscore) (1.4.2)
$s^-(x)$	Inscore of x (1.4.2)
$\zeta(x)$	Minimal gain score of x (6.4.5)
$\eta(x)$	Minimal gain associated to x (6.4.5)

Solutions, general

CV_S	Copeland value of S (9.1.1)
$D(S, T)$	Contestation relation on T associated to S (7)
$S_1 \subset S_2$	S_1 is finer than S_2 : $\forall T, S_1(T) \subset S_2(T)$ and $\exists T, S_1(T) \neq S_2(T)$ (2.2)
$S_1 \cap S_2$	S_1 and S_2 intersect : for no T, $S_1(T) \cap S_2(T) = \varnothing$ (2.2)
$S_1 \varnothing S_2$	S_1 and S_2 may not intersect : $\exists T : S_1(T) \cap S_2(T) = \varnothing$ (2.2)
$S_1 o S_2$	Composition of S_1 and S_2, also denoted $S_1(S_2)$ (2.2)
S^k, S^∞	Iterations of S (2.2)
$Sf(T)$	Set of winners associated to the board f (8.1)
S^F	Algebraic solution for the agenda F (8.1.10)

$\widetilde{S}$	Composition-expansion of S (2.5.1)
S^*	Composition-consistent hull of S (2.5.3)
γ^*, γ^{**}	Two weak expansion properties (5.3.7, 5.3.11)

Solutions, specific

B	Banks' set (7.1)
BP	Bipartisan set (6.3)
C	Copeland solution (3.1)
Ma	Markov solution (3.3)
MC	Minimal covering set (5.3)
SL	Slater solution (3.4)
TC	Top-Cycle (1.3.14)
TEQ	Tournament equilibrium set (7.2)
UC	Uncovered set (5.1)
WUC	Weak uncovered set (5.4)

Miscaleneous

$E(T)$	Equilibrium of the mixed game associated to T (6.2)
$E(T, x^*, a)$	Equilibria of the game with gain a on x^* (6.4)
$G(T)$	Pure game associated to T (6.1)
$J(T)$	Mixed game associated to T (6.2)
$J(T, x^*, a)$	Mixed game associated to T with gain a on x^* (6.4)
M	Majority relation (2.1.2)
Φ	Simple agenda (8.3.1)
Ψ	Sophisticated agenda (8.4.1)
$<f, T, a>$	Winner of T for a along f (8.1)

References

Aïzerman, M. and F. Aleskerov (1995) *Theory of Choice*, North-Holland, Amsterdam.

Anderson, T.W. (1984) *An Introduction to Multivariate Statistical Analysis*, Wiley, New York.

Arrow, K.J. (1951) *Social Choice and Individual Values*, Wiley, New York.

Artin, E. (1978) *Geometric Algebra*, Interscience, New York.

Astié-Vidal, A. and V. Dugat (1993) "Autonomous parts and decomposition of regular tournaments" *Discrete Mathematics* 111:27-36.

Astié-Vidal, A. and A. Matteo (1987) "Non simple tournaments : theoretical properties and a polynomial algorithm", in *Lecture notes in Computer Science* n°156, conference AAECC5, Mexico.

Aupetit, B. and C. Genest (1993) "On some useful properties of the Perron eigenvalue of a positive reciprocal matrix in the context of the analytic hierarchy process" *European Journal of Operation Research* 70:263-268.

Banks, J. (1985) "Sophisticated voting outcomes and agenda control" *Social Choice and Welfare* 2:295-306.

Banks, J., G. Bordes and M. LeBreton (1991) "Covering relations, closest orderings and hamiltonian bypaths in tournaments" *Social Choice and Welfare* 8:355-363.

Barthélemy, J.-P. (1979) "Caractérisation axiomatique de la distance de la différence symétrique entre les relations binaires" *Mathématiques et Sciences Humaines* 67:85-113.

Barthélemy, J.-P. (1990) "Intransitivities of preferences. Lexicographic shifts and the transitive dimension of oriented graphs" *British Journal of Mathematical and Statistical Psychology* 43:29-37.

Barthélemy, J.-P., A. Guénoche and O. Hudry (1989) "Median linear orders : heuristics and a branch and bound algorithm" *European Journal of Operation Research* 41:313-325.

Barthélémy, J.-P. and B. Monjardet (1981) "The median procedure in cluster analysis and social choice theory" *Mathematical Social Sciences* 1:235-267.

Batteau, P., E. Jacquet-Lagrèze and B. Monjardet, eds. (1981) *Analyse et Agrégation des Preférences*, Economica, Paris.

Berge, C. (1970) *Graphes et hypergraphes*, Dunod, Paris (English translation : *Graphs and Hypergraphs*, North Holland Mathematical Library, vol. 6, 1973).

Berggren, J.L. (1972) "An algebraic characterization of finite symmetric tournaments" *Bulletin of the Australian Mathematical Society* 6:53-59.

Bermond, J.-C. (1972) "Ordres à distance minimum d'un tournoi et graphes partiels sans circuits maximaux" *Mathématiques et Sciences Humaines* 37:5-25.

Bermond, J.-C. and Y. Kondratoff (1976) "Une heuristique pour le calcul de l'indice d'intransitivité d'un tournoi" *Revue d'Automatique, d'Informatique et de Recherche Operationnelle* 10:83-92.

Bezembinder, T.G.G. (1981) "Circularity and consistency in paired comparisons" *British Journal of Mathematical and Statistical Psychology* 34:16-37.

Black, D. (1948) "On the rationale of group decision-making" *Journal of Political Economy* 56:23-24.

Black, D. (1958) *The Theory of Committees and Elections*, Cambridge University Press, Cambridge.

Bordes, G. (1976) "Consistency, rationality and collective choice" *Review of Economic Studies* 43:451-457.

Bordes, G. (1979) "Some more results on consistency, rationality and collective choice", in Laffont, J.-J. (ed.) *Aggregation and Revelation of Preferences*, North-Holland, Amsterdam, pp 175-197.

Bordes, G. (1983) "On the possibility of reasonable consistent majoritarian choice: some positive results" *Journal of Economic Theory* 31:122-132.

Bordes, G. (1988) "Voting games, indifference, and consistent sequential choice rules" *Social Choice and Welfare* 5:31-44.

Bradley, R. and M. Terry (1952) "Rank analysis of incomplete block design. I : the method of paired comparisons" *Biometrika* 39:324-345.

Brams, S.J. (1994) "Voting procedures", in R. Aumann and S. Hart (eds.) *Handbook of Game Theory with Economic Applications, vol II,* Elsevier Science, Amsterdam, pp 1055-1089.

Camion, P. (1959) "Chemins et circuits hamiltoniens des graphes complets" *Comptes Rendus de l'Académie des Sciences de Paris (A)* 249:2151-2152.

de Cani, J.-S. (1969) "Maximum likehood paired comparison ranking by linear programming" *Biometrika* 3:537-545.

Charon-Fournier, I., A. Germa and O. Hudry (1992a) "Encadrement de l'indice de Slater d'un tournoi à l'aide de ses scores" *Mathématiques, Informatique et Sciences Humaines* 118:53-68.

Charon-Fournier, I., A. Germa and O. Hudry (1992b) "Utilisation des scores dans des méthodes exactes déterminant les ordres médians des tournois" *Mathématiques, Informatique et Sciences Humaines* 119:53-74.

Charon-Fournier, I., O. Hudry and F. Woigard (1996) "Ordres médians et ordres de Slater des tournois" *Mathématiques, Informatique et Sciences Humaines* 133:23-56.

Chebotarev, P. Yu. and E. Shamis (1995) "Incomplete preferences and indirect scores" preprint of the Institute of Control Sciences, Moscow.

Chen, R. and F. Hwang (1988) "Stronger players win more balanced knockout tournaments" *Graphs and Combinatorics* 4:95-99.

Chung, F. and F. Hwang (1978) "Do stronger players win more knockout tournaments?" *Journal of the American Statistical Association* 73:593-596.

Copeland, A. (1951) "A reasonable social welfare function", University of Michigan seminar on the applications of mathematics to social sciences.

Coughlan, P. and M. LeBreton (1994) "A social choice function implementable via backward induction with values in the ultimate uncovered set" manuscript.

Daniels, H.E. (1969) "Round Robin Tournament scores" *Biometrika* 56:295-299.

David, H. (1963) *The Method of Paired Comparisons*, Griffin's statistical monographs and courses, Charles Griffins, London.

Davidson, R.R. and P.H. Farquhar (1976) "A bibliography on the method of paired comparisons" *Biometrics* 32:241-252.

Deb, R. (1976) "On constructing generalized voting paradoxes" *Review of Economic Studies* 43:347-351.

Deb, R. (1977) "On Schwartz's rule" *Journal of Economic Theory* 16:103-110.

Downs, A. (1957) *An Economic Theory of Democracy*, Harper, New-York.

Dugat, V. (1990) *Décomposition de tournois réguliers : théorie et application aux algorithmes de tests d'isomorphisme.* Thèse de doctorat de l'université Paul Sabatier de Toulouse.

Duggan, J. and M. LeBreton (1996) "Dutta's minimal covering set and Shapley's saddles" *Journal of Economic Theory* 70:257-265.

Dummet, M. (1984) *Voting Procedures*, Clarendon Press, Oxford.

Dutta, B. (1988) "Covering sets and a new Condorcet choice correspondence" *Journal of Economic Theory* 44:63-80.

Dutta, B. (1990) "On the tournament equilibrium set" *Social Choice and Welfare* 7:381-383.

Dutta, B. and A. Sen (1993) "Implementing generalized Condorcet social choice functions via backward induction" *Social Choice and Welfare* 10:149-160.

Erdös, P., E. Fried, A. Hajnal and E.C. Milner (1972) "Some remarks on simple tournaments" *Algebra Universalis* 2:238-245.

Erdös, P. and L. Moser (1964) "On the representation of directed graphs as unions of orderings" *Publications of the Mathematical Institute of the Hungarian Academy of Science* 9:125-132.

Farqharson, R. (1969) *Theory of Voting*, Yale University Press, New Haven.

Felsenthal, D.S. and M. Machover (1992) "After two centuries, should Condorcet's voting procedure be implemented ? " *Behavioral Science* 37:250-274.

Ferejohn, J.A., R.D. McKelvey and E.W. Packel (1984) "Limiting distributions for continuous state Markov models" *Social Choice and Welfare* 1:45-67.

Fishburn, P.C. (1973) *The Theory of Social Choice*, Princeton University press, Princeton, NJ.

Fishburn, P.C. (1974) "Social choice functions" *SIAM Review* 16:63-90.

Fishburn, P.C. (1977) "Condorcet social choice functions" *SIAM Journal of Applied Mathematics* 33:469-489.

Fishburn, P.C. (1991) "Non-transitive preferences in decision theory" *Journal of Risk and Uncertainty* 4:113-134.

Fishburn, P.C. and W.V. Gehrlein (1982) "Majority efficiencies for simple voting procedures" *Theory and Decision* 14:141-153.

Fisher, D. and J. Ryan (1992) "Optimal strategies for a generalized 'Scissors, Paper and Stone' game" *American Mathematical Monthly* 99:935-942.

Fisher, D. and J. Ryan (1995a) "Tournament games and positive tournaments" *Journal of Graph Theory* 19:217-236.

Fisher, D. and J. Ryan (1995b) "Condorcet voting and tournament games" *Linear Algebra and its Applications* 217:87-100.

Fisher, D. and J. Ryan (1995c) "Probabilities within optimal strategies for tournament games" *Discrete Applied Mathematics* 56:87-91.

Ford, L.R. Jr (1957) "Solution of a ranking problem from binary comparisons" *American Mathematical Monthly* 64:28-33

Fried, E. (1970) "On homogeneous tournaments", in *Combinatorial Theory and its Applications, vol. II* (P. Erdös et al. eds.) North-Holland, Amsterdam, pp 467-476.

Fried, E. and H. Laksar (1971) "Simple tournaments" *Notices of the American Mathematical Society* 18:395.

Fulkerson, D.R. (1965) "Upsets in round robin tournaments" *Canadian Journal of Mathematics* 17:957-969.

Genest, C., F. Lapointe and S. Drury (1993) "On a proposal of Jensen for the analysis of ordinal pairwise preferences using Saaty's eigenvector scaling method" *Journal of Mathematical Psychology* 37:575-610.

Genest, C. and L.-P. Rivest (1994) "A statistical look at Saaty's method of estimating pairwise preferences expressed on a ratio scale" *Journal of Mathematical Psychology* 38:477-496.

Gibbard, A. (1973) "Manipulation of voting schemes: a general result" *Econometrica* 41:587-601.

Golden, B., E. Wasil and P. Harker (1989) *The Analytic Hierarcy Process: Applications and Studies*, Springer-Verlag, New York.

Good, I. J. (1971) "A note on Condorcet sets" *Public Choice* 10:97-101.

Greub, W. (1981) *Linear Algebra*, fourth edition: Graduate Texts in Mathematics vol.23, Springer-Verlag, Berlin.

Guénoche, A., B. Vanderputte-Riboud and J.-B. Denis (1994) "Selecting varieties using a series of trials and a combinatorial ordering method" *Agronomie* 14:363-375.

Habib, M. (1981) *Substitution des structures combinatoires, théorie et algorithmes,* thèse d'état, université Pierre et Marie Curie, Paris.

Harary, F. and L. Moser (1966) "The theory of round-robin tournaments" *American Mathematical Monthly* 73:231-246.

Hardouin Duparc, J. (1975) "Quelques résultats sur l'indice de transitivité de certains tournois" *Mathématiques et Sciences Humaines* 51:35-41.

Hartigan, J. (1966) "Probabilistic completion of a knock-out tournament" *Annals of Mathematical Statistics* 37:495-503.

Henriet, D. (1985) "The Copeland choice function: an axiomatic characterization" *Social Choice and Welfare* 2:49-63.

Hollard, G. and M. Le Breton (1995) "Logrolling and a McGarvey theorem for separable tournaments" Document de travail du GREQAM, Marseille.

Hudry, O. (1985) *Recherche d'ordres médians : complexite, algorithmique et problèmes combinatoires,* Thèse de l'Ecole Nationale des Télécommunications, Paris.

Hwang, F. (1982) "An anomaly in knockout tournaments" *Congressus Numerantium* 35:379-386.

Hwang, F. and F. Hsuan (1980) "Stronger players win more knockout tournaments on average" *Commun. Stat. Theory Methods* A9:107-113.

Hwang, F., L. Zongzhen and Y. Yao (1991) "Knock-out tournaments with diluted Bradley-Terry preferences scheme" *Journal of Statistical Planning and Inferences* 28:99-106.

Imrich, W. and J. Nesetril (1992) "Simple tournaments and sharply transitive groups" *Discrete Mathematics* 108:159-165.

Inada, K. (1969) "On the simple majority decision rule" *Econometrica* 37:490-506.

Israel, R. (1982) "Stronger players need not win more knockout tournaments" *Journal of the American Statistical Association* 76:950-951.

Jacquet-Lagrèze, E. (1969) "L'agrégation des opinions individuelles" *Informatique et Sciences Humaines* 4:1-21.

Jech, T. (1983) "The ranking of incomplete tournaments : a mathematician's guide to popular sports" *American Mathematical Monthly* 90:246-266.

Jech, T. (1989) "A quantitative theory of preferences : some results on transition functions" *Social Choice and Welfare* 6:301-314.

Jensen, R. (1984) "An alternative scaling method for priorities in hierarchical structures" *Journal of Mathematical Psychology* 28:317-332.

Jensen, R. (1986) "Comparison of consensus methods for priority ranking problems" *Decision Sciences* 17:195-211.

Jing, H. and L. Weixuan (1987) "Toppling kings by introducing new kings" *Journal of Graph Theory* 11:7-11.

Johnson, C., W. Beine and T. Wang (1979) "Right-left asymmetry in an eigenvector ranking procedure" *Journal of Mathematical Psychology* 19:61-64.

Junger, M. (1985) *Polyhedral Combinatorics and the Acyclic Subdigraph Problem,* Heldermann Verlag, Berlin.

Katz, L. (1953) "A new status index derived from sociometrics analysis" *Psychometrica* 18:39-43.

Keener, J. (1993) "The Perron-Frobenius theorem and the ranking of football teams" *SIAM Review* 35-1:80-93.

Kemeny, J. (1959) "Mathematics without numbers" *Daedalus* 88:577-591.

Kemeny, J. and J. Snell (1962) *Mathematical Models in the Social Sciences,* Gin and co., New York.

Kendall, M.G. (1955) "Further contributions to the theory of paired comparisons" *Biometrics* 11:43-62.

Kendall, M.G. (1970) *Rank Correlation Methods,* 4th edition, Griffin, London.

Kendall, M.G. and B.B. Smith (1940) "On the method of paired comparisons" *Biometrika* 31:324-345.

Laffond, G. and J. Lainé (1994) "Weak covering relations" *Theory and Decision* 37:245-265.

Laffond, G., J. Lainé and J.-F. Laslier (1996) "Composition-consistent tournament solutions and social choice functions" *Social Choice and Welfare* 13:75-93.

Laffond, G. and J.-F. Laslier (1991) "Slater's winners of a tournament may not be in the Banks' set" *Social Choice and Welfare* 8:355-363.

Laffond, G., J.-F. Laslier and M. LeBreton (1991) "A game-theoretical method for ranking the participants in a tournament", Document de travail du Conservatoire National des Arts et Métiers, Paris.

Laffond, G., J.-F. Laslier and M. LeBreton (1993a) "The Bipartisan set of a tournament game" *Games and Economic Behavior* 5:182-201.

Laffond, G., J.-F. Laslier and M. LeBreton (1993b) "More on the Tournament Equilibrium set" *Mathématiques, Informatique et Sciences Humaines* 123:37-43.

Laffond, G., J.-F. Laslier and M. LeBreton (1994a) "Social Choice mediators" *American Economic Review (proc.)* 84:448-453.

Laffond, G., J.-F. Laslier and M. LeBreton (1994b) "The Copeland measure of Condorcet choice functions" *Discrete Applied Mathematics* 55:273-279.

Laffond, G., J.-F. Laslier and M. LeBreton (1995a) "Condorcet choice correspondences: a set-theoretical comparison" *Mathematical Social Sciences* 30:23-36.

Laffond, G., J.-F. Laslier and M. LeBreton (1995b) "*k*-player additive extension of two-player games with an application to the Borda electoral competition game", mimeo.

Laffond, G., J.-F. Laslier and M. LeBreton (1996) "A theorem on symmetric, two-player, zero-sum games" *Journal of Economic Theory* forthcoming.

Landau, H.G. (1953) "On dominance relations and the structure of animal society: III. The condition for a score sequence" *Bulletin of Mathematical Biophysics* 15:143-148.

Laslier, J.-F. (1992) *Solutions de Tournois,* Habilitation à diriger les recherches en Economie, Université de Cergy-Pontoise.

Laslier, J.-F. (1996a) "Multivariate description of comparison matrices" *Multicriteria Decision Analysis* 5:112-126.

Laslier, J.-F. (1996b) "Solutions de tournois : un spicilège" *Mathématiques, Informatique et Sciences Humaines* 133:7-22.

Laslier, J.-F. (1996c) " Rank-based choice correspondences" *Economic Letters* 52:279-286.

Le Breton, M. (1996) Personal communication.

Lepelley, D. (1993) "On the probability of electing the Condorcet loser" *Mathematical Social Sciences* 25:105-116.

Levchenkov, V. (1992) "Social Choice Theory: a new insight" preprint of the Institute for System Analysis, Moscow.

Levchenkov, V. (1995a) "Self-consistent rule for group choice. I: Axiomatic approach" Conservatoire National des Arts et Métiers, discussion paper #95-3.

Levchenkov, V. (1995b) "Self-consistent rule for group choice. II: Dynamic approach" Conservatoire National des Arts et Métiers, discussion paper #95-4.

Levchenkov, V. (1995c) "Cyclic tournaments: a matching solution" mimeo.

Maas, A. (1993) "A relativized measure of circularity in tournaments" *Mathematical Social Sciences* 26:79-91.

Marcotorchino, J.-F. and P. Michaud (1979) *Optimisation en analyse ordinale des données*, Masson, Paris.

Maurer, S.B. (1980) "The king chicken theorems" *Mathematics Magazine* 53:67-80.

Maybee, J.S. and N.J. Pullman (1990) "Tournament matrices and their generalizations, I" *Linear and Multilinear Algebra* 28:57-70.

McGarvey, D. (1953) "A theorem on the construction of voting paradoxes" *Econometrica* 21:608-610.

Mc Kelvey, R.D. (1979) "General conditions for global intransitivities in a formal voting model" *Econometrica* 47:1085-1112.

Mc Kelvey, R.D. (1986) "Covering, dominance, and institution-free properties of social choice" *American Journal of Political Science* 30:283-314.

McKelvey, R.D. and R.G. Niemi (1978) "A multistage game representation of sophisticated voting for binary procedures" *Journal of Economic Theory* 18:1-22.

McLean, I. and A. Urken (eds.) (1995) *Classics of Social Choice*, The University of Michigan Press, Ann Arbour.

Merril, S. (1988) *Making Multicandidate Elections More Democratic*, Princeton University press, Princeton NJ.

Merlin, V. and D.G. Saari (1995a) "The Copeland Method I: Relationships and the dictionary" mimeo.

Merlin, V. and D.G. Saari (1995b) "The Copeland Method II: Manipulation, monotonicity and paradoxes" mimeo.

Miller, N.R. (1977) "Graph-theoretical approaches to the theory of voting" *American Journal of Political Science* 21:769-803.

Miller, N.R. (1980) "A new solution set for tournaments and majority voting: further graph-theoretical approaches to the theory of voting" *American Journal of Political Science* 24:68-96.

Miller, N.R. (1983) "The covering relation in tournaments: two corrections" *American Journal of Political Science* 27:382-385. (Erratum in 28:434.)

Miller, N.R., B. Grofman and S.L. Feld (1990) "The structure of the Banks set" *Public Choice* 66:243-251.

Möhring, R.H. and F.J. Radermacher (1984) "Substitution decomposition for discrete structures and connections with combinatorial optimization" *Annals of Discrete Mathematics* 19:257-356.

Monjardet, B. (1973) "Tournois et ordre médians pour une opinion" *Mathématiques et Sciences Humaines* 43:55-70.

Monjardet, B. (1979) "Relations à éloignement minimum de relations binaires, note bibliographique" *Mathématiques et Sciences Humaines* 67:115-122.

Monjardet, B. (1990) "Sur diverses formes de la "Règle de Condorcet" d'agrégation des préférences" *Mathématiques, Informatique et Sciences Humaines* 111:61-71.

Monsuur, H. and T. Storken (1996) "Measuring inconsistency" mimeo.

Moon, J.W. (1968) *Topics on Tournaments,* Holt, Rinehart and Winston, New York.

Moon, J.W. (1972) "Embedding tournaments in simple tournaments" *Discrete Mathematics* 2:389-395.

Moon, J.W. and N. J. Pullman (1967) "On the power of tournament matrices" *Journal of Combinatorial Theory* 3:1-9.

Moon, J.W. and N. J. Pullman (1970) "On generalized tournament matrices" *SIAM Review* 12:384-399.

Moulin, H. (1979) "Dominance-solvable voting schemes" *Econometrica* 47:1337-1351.

Moulin, H. (1981) *Théorie des Jeux pour l'Economie et la Politique,* Hermann, Paris (English translation: *Game Theory for the Social Sciences,* New York University Press, 1982).

Moulin, H. (1983) *The Strategy of Social Choice,* Advanced textbooks in Economics, vol. 18, North-Holland, Amsterdam.

Moulin, H. (1985) "Choice functions over a finite set : a summary" *Social Choice and Welfare* 2:147-160.

Moulin, H. (1986) "Choosing from a tournament" *Social Choice and Welfare* 3:271-291.

Moulin, H. (1988) *Axioms of Cooperative Decision Making,* Cambridge University Press, Cambridge.

Muller, V., J. Nesetril and J. Pelant (1975) "Either tournaments or algebras ?" *Discrete Mathematics* 11:37-66.

Myerson, R. (1991) *Game Theory: Analysis of Conflict,* Harvard University Press, Cambridge Mass.

Myerson, R. (1995) "Axiomatic derivation of scoring rules without the ordering assumption" *Social Choice and Welfare* 12:59-74.

Ng, Y.K. (1989) "Individual irrationality and social welfare" *Social Choice and Welfare* 6:87-101.

Nurmi, H. (1987) *Comparing Voting Systems,* Kluwer Academic Press, Boston.

Nurmi, H. (1992) "An assesment of voting system simulations" *Public Choice* 73:459-487.

Ordeshook, P.C. (1986) *Game Theory and Political Theory, an Introduction,* Cambridge University Press, Cambridge.

Ordeshook, P.C. (1992) *A Political Theory Primer,* Routledge, London.

Ordeshook, P.C. and T. Schwartz (1987) "Agenda control and the control of political outcomes" *American Political Science Review* 81:179-199.

Owen, G. (1982) *Game Theory* (second edition), Academic Press, New York.

Perny, P. (1994) "Defining fuzzy covering relations for multiple criteria decision aid" *Proc. Fifth Int. Conf. IPMU, Paris July 1994, vol II,* pp 1019-1025.

Reid, K.B. (1982) "Every vertex is a king" *Discrete Mathematics* 38:93-98.

Reid, K.B. (1991a) "Majority tournaments: sincere and sophistigated voting decisions under amendment procedure" *Mathematical Social Sciences* 21:1-19.

Reid, K.B. (1991b) "The relationship between two algorithms for decisions via sophistigated majority voting with an agenda" *Discrete Applied Mathematics* 31:23-28.

Reid, K.B. and L.W. Beineke (1978) "Tournaments", in L.W. Beineke and R.J. Wilson (eds.) *Selected Topics in Graph Theory,* Academic Press, pp 169-204.

Remage, R. and W.A. Thompson (1966) "Maximum likehood paired comparison rankings" *Biometrika* 53:143-149.

Roubens, M. and Ph. Vincke (1985) *Preference Modelling*, Springer-Verlag, Berlin.

Roy, B. and D. Bouyssou (1992) *Aide multicritère à la décision : méthodes et cas*, Economica, Paris.

Rubinstein, A. (1980) "Ranking the participants in a tournament" *SIAM Journal of Applied Mathematics* 38:108-111.

Ryser, H.J. (1964) *Matrices of Zeros and Ones in Combinatorial Mathematics, Recent Advances in Matrix Theory*, Madison, Wis.

Saary, D. G. (1994) *Geometry of Voting*, Springer-Verlag, Berlin.

Saaty, T. (1977) A scaling method for priorities in hierarchical structures" *Journal of Mathematical Psychology* 15:234-281.

Saaty, T. (1980) *The Analytic Hierarchy Process*, McGraw-Hill, New York.

Satterwaite, M.A. (1975) "Strategyproofness and Arrow's conditions: existence and correspondences for voting procedures and social welfare functions" *Journal of Economic Theory* 10:187-217.

Schwartz, T. (1972) "Rationality and the myth of maximum" *Noûs* 6:97-117.

Schwartz, T. (1977) "Collective choice, separation of issues and vote trading" *American Political Science Review* 71:999-1010.

Schwartz, T. (1990) "Cyclic tournaments and cooperative majority voting : a solution" *Social Choice and Welfare* 7:19-29.

Seber, G.A. (1984) *Multivariate Observations*, Wiley series in probability and statistics, John Wiley and sons, New York.

Sen, A.K. (1971) "Choice functions and revealed preferences" *Review of Economic Studies* 38:307-317.

Shapley, L. (1964) "Some topics in two-person games", in M. Dresher, L. Shapley and A. Tucker (eds.) *Advances in Game Theory*, Annals of Mathematics Studies 52, Princeton University Press, pp 1-28.

Shepsle, K. and B. Weingast (1984) "Uncovered sets and sophisticated voting outcomes with implications for agenda institution" *American Journal of Political Science* 28:49-74.

Slater, P. (1961) "Inconsistencies in a schedule of paired comparisons" *Biometrika* 48:303.312.

Smith, J. (1973) "Aggregation of preferences with a variable electorate" *Econometrica* 41:1027-1041.

Stearns, R. (1959) "The voting problem" *American Mathematical Monthly*, 66:761-763.

Thompson, G.L. (1958) *Lectures on Game Theory, Markov Chains and Related Topics,* Sandia Corporation Monograph SCR-11.

Thompson, W.A. and R. Remage (1964) "Rankings from paired comparisons" *Annals of Mathematics and Statistics* 35:739-747.

Tideman, T. N. (1987) "Independence of clones as a criterion for voting rules" *Social Choice and Welfare* 4:185-206.

Tideman, T. N. and T.M. Zavist (1989) "Complete independence of clones in the ranked pairs rules" *Social Choice and Welfare* 6:167-173.

Tversky, A. (1969) "Intransitivity of preferences" *Psychological Review,* 76:31-48.

Ushakov, I.A. (1976) "The problem of choosing the preferred element: an application to sports games", in R.E. Machol, S.P. Ladany and D.G. Morrison (eds.), *Management Science in Sports* North-Holland, Amsterdam, pp 153-161.

Vargas, L. and R. Whittaver (guest eds.) (1990) "Decison making by the analytic hierarchy process: Theory and applications" *European Journal of Operational Research* 48:1.

Varlet, J.-C. (1976) "Convexity in tournaments" *Bulletin de la Société Royale des Sciences de Liège* 11-12:570-586.

Volle, M. (1985) *Analyse des données,* Economica, Paris.

Wei, T. (1952) *The Algebraic Foundation of Ranking Theory,* Ph. D. Thesis, Cambridge University.

Young, H.P. (1974) "An axiomatization of Borda's rule" *Journal of Economic Theory* 9:43-52.

Young, H.P. (1975) "Social choice scoring functions" *SIAM Journal of Applied Mathematics* 28:824-838.

Young, H.P. (1988) "Condorcet's theory of voting" *American Political Science Review* 82:1231-1244.

Young, H.P. and A. Levenglick (1978) "A consistent extension of Condorcet's election principle" *SIAM Journal of Applied Mathematics* 35:285-300.

Zermelo, E. (1929) "Die Berechnung der Turnier-Ergebnisse als ein Maximalproblem der Wahrscheinlichkeitsrechnung *Mathematische Zeitschrift* 29:436-460.

Economic Theory

Official journal of the Society for the Advancement of Economic Theory

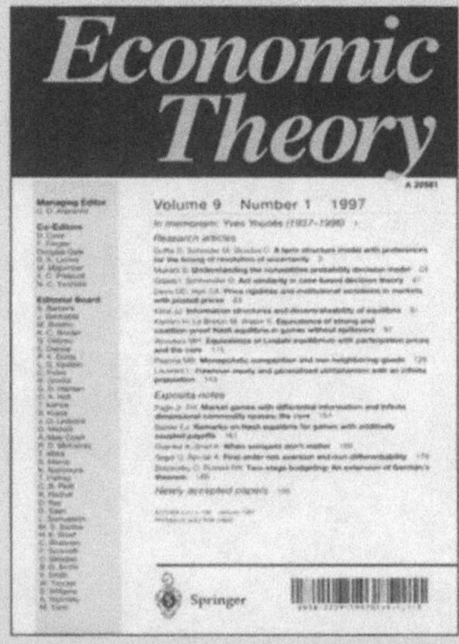

Managing Editor: **C.D. Aliprantis**, IUPUI, Indianapolis, USA

A selection of forthcoming papers in 1997

D. G. Saari:
The generic existence of a core for q-rules.

C. Gilles, S.F. LeRoy:
Bubbles as payoffs at infinity.

J. Banks, M. Olson, D. Porter:
An experimental analysis of the bandit problem.

M. Magill, M. Quinzii:
Which improves welfare more: nominal or indexed bond?

E. Bond, T. Gresik:
Competition between asymmetrically informed principals.

G. Hahn, N.C. Yannelis:
Efficiency and incentive compatibility in differential information economies.

A. Goenka, K. Shell:
On the robustness of sunspot equilibria.

T.J. Sargent, B.D. Smith:
Coinage, debasements and Gresham's laws.

GOING ONLINE

Economic Theory is included in the programme **"LINK – Springer Print Journals Go Electronic".**
Electronic edition:
ISSN 1432-0479

For more information, please visit us on **http://link.springer.de**
or by fax: **+49-6221-487-288.**
In 1997, access to the electronic edition is **free for subscribers** to the printed journal.

ISSN 0938-2259 Titel No. 199

Subscription information 1997
Volume 9+10 (6 issues)
DM 980,–*

Ask for your sample copy

* suggested list price, plus carriage charges.
In EU countries the local VAT is effective.

Please order by
Fax: +49-30-827 87-448
e-mail: subscriptions@springer.de
or through your bookseller

Springer

Springer-Verlag, P. O. Box 31 13 40, D-10643 Berlin, Germany.

Jak.4021/MNTZ/E/1